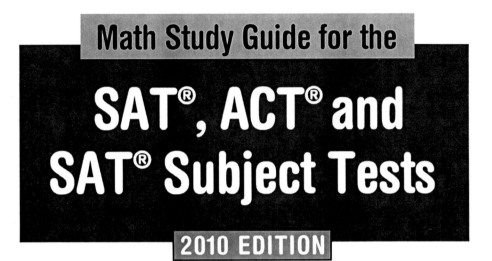

Math Study Guide for the

SAT®, ACT® and SAT® Subject Tests

2010 EDITION

Created by
Richard F. Corn

√mpb

Math Prep Books
A Wyatt-MacKenzie Imprint

Math Study Guide for the SAT®, ACT®, and SAT® Subject Tests — 2010 Edition

Created by Richard F. Corn

ISBN: 978-0-9743832-9-3

Library of Congress Control Number: 2009928158

Errata for this book may be found at
http://www.mathprepbooks.com/errata.html

Published by Math Prep Books, A Wyatt-MacKenzie Imprint
www.wymacpublishing.com

Math Prep Books
A Wyatt-MacKenzie Imprint

TABLE OF CONTENTS

1. INTRODUCTION

Buying this one book will enable you to prepare for the math sections of any of the standardized tests for college admission, either the SAT®, the ACT®, or the math SAT® subject tests (formerly called the SAT II subject tests). You can prepare for any test or combination of tests, in any order, without starting from scratch each time. Material needed for each of the tests is clearly identified, so that you can study only the math needed for a particular test.

The material in this book has been field tested over and over again. It is the product of many hours of helping students prepare for the math sections of the SAT® and ACT®, and for the math SAT® subject tests. The lessons and problems in this book are real, in that my students have used them, asked questions about them, and pointed out improvements to be made. This 2010 edition is an improved version of the book, and most of those improvements are a direct result of student comments. In addition, I track the test scores obtained by students who I have tutored, and these materials have helped them achieve excellent results.

As an active math tutor, I know that students routinely take both the SAT® and the ACT® exams (and sometimes take a math SAT® subject test as well). In my practice, I found that it is better to gather all the preparation material in a single book rather than start from scratch with a new book for each test.

The content of this book is based on these primary sources:

- The math syllabus published in the "preparation booklets" by the College Board and by ACT, Inc. These booklets can sometimes be found in your school's guidance office. The Internet links to these booklets are given below.

- The math chapters and practice tests in The Official SAT Study Guide published by the College Board, The Real ACT Prep Guide published by ACT, Inc., and The Official SAT Subject Tests in Mathematics Levels 1 and 2 Study Guide published by the College Board. These books are excellent sources of practice tests and are available at most book sellers.

How to use this book

This book enables you to build a solid math foundation for the standardized tests for college admission. Schedule your time to do several units each week, and ideally complete all of the units before taking the test. In addition, you should complete a series of practice tests to develop test-taking skills. Test-taking skills are an essential component of preparation and must not be overlooked.

To prepare for the SAT® only, study chapters 1-6 of this book. To prepare for the ACT® or one of the SAT® subject tests, study chapters 1-7 of this book. Chapters and units within chapters may be studied in any order.

SAT is a registered trademark of the College Board and ACT is a registered trademark of ACT, Inc. Neither organization was involved in the production of, and does not endorse, this book.

Practice tests

This book does not contain practice tests because in my opinion, the best sources for practice tests are the test makers themselves. Each offers a free "preparation booklet" that can be downloaded from their web site, and each preparation booklet contains a practice test along with other useful information. Links to the "preparation booklets" are below. Each test maker also publishes a book of practice tests (available through most book sellers) and these are given below as well:

A free practice SAT® may be found at http://www.collegeboard.com/student/testing/sat/prep_one/test.html?BannerID=pretest&AffiliateID=PREPCENTER . The College Board also publishes a book of eight practice tests, called The Official SAT Study Guide.

A free practice ACT® may be found at: http://www.act.org/aap/pdf/preparing.pdf . ACT, Inc. also publishes a book of three practice tests, called The Real ACT Prep Guide.

I have written a book that is a companion to the SAT® and ACT® books referenced above. It is called Math Secrets for the SAT® and ACT®. It may be found through most online retailers. Using examples from official published practice tests, it provides students with test-taking tips.

Free practice questions for each of the SAT® subject tests may be found at

http://www.collegeboard.com/prod_downloads/sat/sat-subject-test-preparation-booklet.pdf . Official practice tests for the math SAT® subject tests may be found in a book called The Official SAT Subject Tests in Mathematics Levels 1 & 2 Study Guide. It is published by the College Board, and it contains two level 1 tests and two level 2 tests.

What follows are lessons and homework problems for the math sections of the SAT®, ACT® and SAT® subject tests. These should be supplemented by lots of time spent on practice tests. The combination of math review and practice tests is the best way to prepare. If you start your preparation early and do parts of this book every week in combination with practice tests in the weeks leading up to the big day, your confidence and scores should improve.

Feedback and Errata

I am always interested in receiving feedback on this book, in order to make it better for the next group of students who use it. Please send feedback, including any errors you may have discovered, to mathprepbooks@gmail.com.

Errata for this book may be found at http://www.mathprepbooks.com/errata.html. Errors will be posted promptly, as they are discovered and verified.

2. PRE-ALGEBRA

This book begins by covering the math that you were supposed to have mastered in middle school. Do not skip this chapter even though you think this is "easy math" because

- You may not have mastered the material at the time you were in middle school
- You saw this so many years ago that you forgot some of it
- It takes practice to get used to the way this material is presented on standardized tests.

Just about every student needs a refresher on pre-algebra, for any combination of the reasons above.

Unit 2.1 Integers, primes and digits

For the purposes of the test, you can think of an integer as any number with all zeros to the right of the decimal. Below are examples:

3   is an integer because it can be written as 3.0000000000

-2   is an integer because it can be written as -2.000000000

0   is an integer because it can be written as 0.0000000000

1/2   is not an integer because it can be written as 0.50000000. There are not all zeros to the right of the decimal.

$\sqrt{2}$ is not an integer because it can be written as 1.4142135.

π is not an integer because it can be written as 3.1415926.

Formally, we can write the set of integers as $\{...,-3,-2,-1,0,1,2,3,...\}$.

Substitution rules

Substitution is a technique that is commonly used to solve problems on standardized tests. When certain words or phrases appear in a problem, it is often helpful to substitute specific values in order to solve the problem. Below is a table of popular phrases and the values that should be substituted.

Phrase	Substitute
Integer	0
Positive integer	1
Negative integer	-1
Even integer	0
Odd integer	1
Consecutive integers	0, 1, 2, etc.
Consecutive even integers	0, 2, 4, etc.
Consecutive odd integers	1, 3, 5, etc.

There will also be problems on the test where substitution is not appropriate, and where algebra is needed. In those problems, you will use algebraic expressions for the phrases above.

Phrase	Substitute
Integer	n
Even integer	2n
Odd integer	2n+1
Consecutive integers	n, n+1, n+2, etc.
Consecutive even integers	n, n+2, n+4, etc. or 2n, 2n+2, 2n+4, etc.
Consecutive odd integers	n, n+2, n+4, etc. or 2n+1, 2n+3, 2n+5

Primes

A prime number is a positive integer that is divisible by itself and 1. Note that 1 is **not** a prime number. The smallest prime number is 2 and all the other prime numbers are odd. Two is the smallest prime and the only even prime number.

The prime numbers are: {2, 3, 5, 7, 11, 13, 17, 19, 23, ...}

Note that 9 is not prime (it is divisible by 3), 15 is not prime (it is divisible by 3 and 5), and 21 is not prime (it is divisible by 3 and 7). A good exercise is to write down the first 20 or so prime numbers.

Problems on the test will use phrases that involve the word "prime." When you see these phrases, it is often helpful to substitute specific values in order to solve the problem. Below is a table of some popular phrases and the values to be substituted.

Phrase	Substitute
Prime number	2
Even prime	2
Odd prime	3
Consecutive primes	2, 3, 5, etc.

A concept related to prime numbers is called prime factorization. The idea is that any integer (that is not itself a prime) can be expressed as the product of primes. It can be very useful to construct a factor tree for any given number. Below is the factor tree for 210.

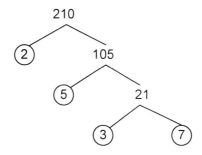

Notice that 210 is an even number. So we divide by two and get 105. The two is circled because two is prime. We notice that 105 is divisible by 5. So we divide by five and get 21. Five is circled because it is prime. The number 21 is divisible by three and seven, which are circled because they are prime.

The prime factorization of 210 is $210 = 2 \cdot 3 \cdot 5 \cdot 7$. The number 210 can be expressed as a product of primes because 210 is not itself prime.

Prime factorization comes in handy later, when adding or subtracting fractions (unit 2.2), finding the common factor (unit 3.4), and factoring polynomials (unit 3.5).

Digits and place value

You may be wondering what this topic is doing in here. Well, the SAT® syllabus has digits and the ACT® syllabus has place value, that's why. Questions on this topic do occur.

Consider the number

$$2345.907$$

It consists of seven digits, ranging from zero through 9. In fact, the set of all possible values for any particular digit appearing in any position is $\{0, 1, 2, 3, 4, 5, 6, 7, 8, 9\}$. The place values are:

$2 =$ the thousands digit

$3 =$ the hundreds digit

$4 =$ the tens digit

$5 =$ the ones or unit digit

$9 =$ the tenths digit

$0 =$ the hundredths digit

$7 =$ the thousandths digit

A common mistake is to choose 3 for hundredths digit instead of 0, or choose 9 for the tens digit instead of 4. This is simple stuff, but it is easy to make a mistake.

Problems on integers, primes and digits (unit 2.1)

1. Which of the following numbers is prime?

    (A)   1
    (B)   55
    (C)   71
    (D)   1617
    (E)   3334

2. If x is a positive integer, what is the smallest possible value of $\dfrac{x+7}{4}$?

    (A)   0
    (B)   1
    (C)   2
    (D)   3
    (E)   4

3. The sum of three consecutive odd integers is 171. What is the value of the largest integer?

    (A)   27
    (B)   31
    (C)   55
    (D)   59
    (E)   171

4. For the number 4768.325, what is the sum of the tens digit and the tenths digit?

    (A)   8
    (B)   9
    (C)   10
    (D)   11
    (E)   12

5. If the sum of two prime numbers is odd, one of the numbers must be:

    (A)   0
    (B)   1
    (C)   2
    (D)   3
    (E)   4

6. If m is an even integer and n is an odd integer, which expression must be odd?

    (A) $2m + n - 1$
    (B) $m \cdot n$
    (C) $m + 2n$
    (D) $m - n$
    (E) $m + 2$

7. The sum of seven consecutive integers is zero. What is the value of the smallest integer?

    (A)   -3
    (B)   -2
    (C)   -1
    (D)   0
    (E)   1

8. The number 509 is prime. What is the next largest prime?

    (A)   510
    (B)   511
    (C)   515
    (D)   519
    (E)   521

9. The sum of three integers is odd. Which of the statements below must be false?

    I.   All three integers are odd.

    II.   Only two integers are odd

    III. Only one integer is odd

    (A)   I only
    (B)   II only
    (C)   III only
    (D)   I and III only
    (E)   I, II, and III

10. If the average of four consecutive odd integers is 20, what is the largest integer?

    (A)   17
    (B)   19
    (C)   20
    (D)   21
    (E)   23

Solutions to problems on integers, primes and digits (unit 2.1)

1. (C)

The factor trees are:

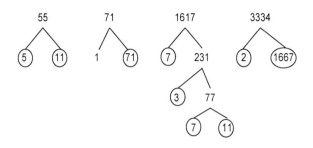

1 is not a prime number.

$55 = 5 \cdot 11$

$1617 = 3 \cdot 7^2 \cdot 11$

$3334 = 2 \cdot 1667$

2. (C)

The smallest positive integer is 1. So the smallest value of $\dfrac{x+7}{4}$ is $\dfrac{1+7}{4} = 2$.

3. (D)

Set up the equation $n + (n+2) + (n+4) = 171$. Simplifying gives $3n + 6 = 171$, or $n = 55$. The largest integer is $n + 4 = 55 + 4 = 59$

4. (B)

The tens digit is six and the tenths digit is three, 6+3=9.

5. (C)

All primes are odd except 2. The sum of two odds is even, so two odds will not work. For the sum to be odd, one of the primes must be even, and the only even prime is 2.

6. (D)

The easiest way to solve this is with substitution. Let $m = 0$ and $n = 1$. Then $m - n = -1$, which is odd. The other choices produce even results.

7. (A)

The quick way to solve this is to realize that half of the integers must be negative and the other half positive. If there are seven integers total, then they must be -3, -2, -1, 0, 1, 2, 3. The smallest integer is -3. This could be solved with algebra instead. Start with the equation

$n + (n+1) + (n+2) + (n+3) + (n+4) + (n+5) + (n+6) = 0$

Simplifying gives $7n + 21 = 0$, or n=-3.

8. (E)

Iterate from 509, skipping the even numbers. 511 is not prime (511/7 = 73). 513 is not prime (513/3 = 171). 515 is not prime (515/5 = 103). 517 is not prime (517/11 = 47). 519 is not prime (519/3 = 173). However 521 is prime.

9. (B)

Statement I may be true because 1+1+1=3. Statement III may be true because 1+0+0=1. However statement II must be false.

10. (E)

$$\frac{n + (n+2) + (n+4) + (n+6)}{4} = 20$$

$4n + 12 = 80$

$4n = 68,\ n = 17$

$n + 6 = 23$

Unit 2.2 Fractions

This unit covers operations on fractions, with some special twists favored on the standardized tests.

Adding and Subtracting

Please remember that **<u>fractions cannot be added or subtracted unless they have the same denominator</u>**, called the common denominator.

Consider a simple case: $\dfrac{1}{2} + \dfrac{2}{7}$. Hopefully you are not thinking that the answer is $\dfrac{1}{2} + \dfrac{2}{7} = \dfrac{1+2}{2+7} = \dfrac{3}{9} = \dfrac{1}{3}$!!!!!
That would be breaking just about every rule in math. (Believe it or not I have seen students do exactly that.)
Remember to first convert each fraction so that both fractions have the same denominator, then add or subtract:

$$\frac{1}{2} + \frac{2}{7} = \left(\frac{1}{2}\right)\left(\frac{7}{7}\right) + \left(\frac{2}{7}\right)\left(\frac{2}{2}\right) = \frac{7}{14} + \frac{4}{14} = \frac{7+4}{14} = \frac{11}{14}$$

In this example the common denominator is 14. Although the product of the denominators can always be used, your work will be a bit simpler if you use the least common denominator (LCD). Consider the example of $\dfrac{7}{15} - \dfrac{2}{20}$. You could use 300 as the common denominator:

$$\frac{7}{15} - \frac{3}{20} = \left(\frac{7}{15}\right)\left(\frac{20}{20}\right) - \left(\frac{3}{20}\right)\left(\frac{15}{15}\right) = \frac{140}{300} - \frac{45}{300} = \frac{140-45}{300} = \frac{95}{300} = \frac{19}{60}$$

The drawback to this approach is that you are working with larger numbers, and you have to do some simplification at the end. On the other hand, with the graphing calculator, simplifying fractions is easy using the Math-Frac keys. The key sequence is: 95, ÷, 300, *math, fract, enter*. So if finding the LCD drives you crazy, you can use this approach together with the calculator.

It is a little easier if you can find the LCD and use that instead. To find the LCD, calculate the prime factorization (see unit 2.1) of each denominator: $15 = 3 \cdot 5$ and $20 = 2^2 \cdot 5$. Next, make a list of all of the prime factors found in either denominator, and then raise each prime factor to its largest power:

$$LCD = 2^2 \cdot 3^1 \cdot 5^1 = 60.$$

Believe it or not, this technique for finding the LCD always works. But if you only have two fractions to deal with, it is much easier to use Math-Num-lcm keys on the graphing calculator. The key sequence is
math − num − lcm − enter −15−, −60−) − enter. Now that we found the LCD is 60, we proceed with the subtraction:

$$\frac{7}{15} - \frac{3}{20} = \left(\frac{7}{15}\right)\left(\frac{4}{4}\right) - \left(\frac{3}{20}\right)\left(\frac{3}{3}\right) = \frac{28}{60} - \frac{9}{60} = \frac{19}{60} .$$

You may be wondering if this LCD stuff is worth all the effort. When you are only dealing with two fractions, it may not be. When you are dealing with several fractions it is far better. There are some problems like that at the end of the unit.

It is also necessary to add and subtract fractions involving variables. A simple example would be $\frac{3}{x} + \frac{y}{2}$. The common denominator would be 2x, and this yields

$$\frac{3}{x} + \frac{y}{2} = \left(\frac{3}{x}\right)\left(\frac{2}{2}\right) + \left(\frac{y}{2}\right)\left(\frac{x}{x}\right) = \frac{6}{2x} + \frac{yx}{2x} = \frac{xy + 6}{2x} .$$

Multiplying

When multiplying fractions, multiply the numerators together to form the new numerator, and multiply the denominators together to form the new denominator. For example,

$$\left(\frac{2}{3}\right)\left(\frac{5}{7}\right) = \frac{2 \cdot 5}{3 \cdot 7} = \frac{10}{21}$$

The same rule applies to fractions with variables:

$$\left(\frac{x}{3y}\right)\left(\frac{4}{z}\right) = \frac{4x}{3yz}$$

Dividing

When two fractions are to be divided, multiply the fraction in the numerator by the reciprocal of the fraction in the denominator. For example,

$$\left(\frac{2}{3}\right) \div \left(\frac{5}{7}\right) = \left(\frac{2}{3}\right)\left(\frac{7}{5}\right) = \frac{14}{15}$$

The same rule applies to fractions with variables:

$$\left(\frac{x}{3y}\right) \div \left(\frac{4}{z}\right) = \left(\frac{x}{3y}\right)\left(\frac{z}{4}\right) = \frac{xz}{12y}$$

Problems on fractions (unit 2.2)

1. What is the value of $\dfrac{7}{10} - \dfrac{5}{12}$?

   (A)   6/11

   (B)   7/24

   (C)   17/60

   (D)   35/120

   (E)   67/60

2. What is the value of $\dfrac{7}{10} + \dfrac{5}{12} - \dfrac{11}{42}$?

   (A)   359/420

   (B)   193/140

   (C)   -1/20

   (D)   11/144

   (E)   118/525

3. Subtract $\dfrac{2}{5x^2} - \dfrac{3}{25xy} = ?$.

   (A) $7/5x^2y$

   (B) $-1/(5x(x-5y))$

   (C) $(10y-3x)/5x^2y$

   (D) $(10y-3x)/25x^2y$

   (E) $7/(x-5y)$

4. Subtract $\dfrac{x}{1-x} - \dfrac{y}{1-y} = ?$

   (A) $(x+y)/(y-x)$

   (B) $(x-y)/(x^2-2xy+y^2)$

   (C) $(x+y-2xy)/(2-x-y)$

   (D) $(y-x)/(x-y)$

   (E) $(x-y)/(xy-x-y+1)$

5. What is the value of $\left(\dfrac{3}{5}\right) \cdot \left(\dfrac{-5}{12}\right)$?

   (A)   -2/17

   (B)   -1/4

   (C)   -36/25

   (D)   -15/17

   (E)   3/17

6. What is the value of $\left(\dfrac{3}{5}\right) \div \left(\dfrac{-5}{12}\right)$?

   (A)   -1/4

   (B)   -36/25

   (C)   -4/1

   (D)   -15/60

   (E)   -2/17

7. Find an expression for $\left(\dfrac{5x^2}{3y}\right) \cdot \left(\dfrac{6y^2}{8x}\right)$

   (A) $5x/4y$

   (B) $20xy/9$

   (C) $4y/5x$

   (D) $20x^3/9y^3$

   (E) $5xy/4$

8. Find an expression for $\left(\dfrac{5x^2}{3y}\right) \div \left(\dfrac{6y^2}{8x}\right)$

   (A) $5xy/4$

   (B) $20xy/9$

   (C) $4/5xy$

   (D) $20x^3/9y^3$

   (E) $5x/4y$

Solutions to problems on fractions (unit 2.2)

1. (C)

It is best to find the LCD.

$10 = 2 \cdot 5$ and $12 = 2^2 \cdot 3$. So the LCD

is $2^2 \cdot 3 \cdot 5 = 60$. $\quad \dfrac{7}{10} - \dfrac{5}{12} = \dfrac{42}{60} - \dfrac{25}{60} = \dfrac{17}{60}$

2. (A)

Again, find the LCD.

$10 = 2 \cdot 5$ and $12 = 2^2 \cdot 3$ and $42 = 2 \cdot 3 \cdot 7$. So

the LCD is $2^2 \cdot 3 \cdot 5 \cdot 7 = 420$.

$\dfrac{7}{10} + \dfrac{5}{12} - \dfrac{11}{42} = \dfrac{7 \cdot 42}{420} + \dfrac{5 \cdot 35}{420} - \dfrac{11 \cdot 10}{420} = \dfrac{359}{420}$

3. (D)

$\dfrac{2}{5x^2} - \dfrac{3}{25xy} = \dfrac{2(5y)}{25x^2y} - \dfrac{3x}{25x^2y} = \dfrac{10y - 3x}{25x^2y}$

4. (E)

$\dfrac{x}{1-x} - \dfrac{y}{1-y} = \dfrac{x(1-y) - y(1-x)}{(1-x)(1-y)} = \dfrac{x-y}{xy - x - y + 1}$

5. (B)

$\left(\dfrac{3}{5}\right)\left(\dfrac{-5}{12}\right) = \dfrac{-3 \cdot 5}{5 \cdot 12} = \dfrac{-15}{60} = \dfrac{-1}{4}$ or

$\left(\dfrac{3}{5}\right)\left(\dfrac{-5}{12}\right) = \left(\dfrac{3}{1}\right)\left(\dfrac{-1}{12}\right) = \left(\dfrac{1}{1}\right)\left(\dfrac{-1}{4}\right) = \dfrac{-1}{4}$

6. (B)

$\left(\dfrac{3}{5}\right) \div \left(\dfrac{-5}{12}\right) = \left(\dfrac{3}{5}\right)\left(\dfrac{12}{-5}\right) = \dfrac{36}{-25} = -\dfrac{36}{25}$

7. (E)

$\left(\dfrac{5x^2}{3y}\right)\left(\dfrac{6y^2}{8x}\right) = \dfrac{5 \cdot 6x^2y^2}{3 \cdot 8xy} = \dfrac{30x^2y^2}{24xy} = \dfrac{5xy}{4}$

8. (D)

$\left(\dfrac{5x^2}{3y}\right) \div \left(\dfrac{6y^2}{8x}\right) = \left(\dfrac{5x^2}{3y}\right)\left(\dfrac{8x}{6y^2}\right) = \dfrac{40x^3}{18y^3} = \dfrac{20x^3}{9y^3}$

Unit 2.3 Mixed numbers and remainders

The standardized tests take a different view of mixed numbers than the view taken in most middle schools.

The main point that is likely to be tested is converting between improper fractions and mixed numbers, with an emphasis on finding the remainder.

Converting improper fractions to mixed numbers

An improper fraction is a fraction where the numerator is larger than the denominator, for example $\dfrac{7}{4}$. The improper fraction is converted to a mixed number by dividing to get a quotient and a remainder.

$$\frac{7}{4} = \frac{4+3}{4} = \frac{4}{4} + \frac{3}{4} = 1 + \frac{3}{4} = 1\frac{3}{4}$$

In this example, the quotient is 1 (the whole number part) and the remainder is 3 (the numerator of the fraction part of the mixed number).

In the real world, you would not do conversions in this way. You would use a calculator. First, divide 4 into 7

$$7 \div 4 = 1.75$$

So now we know that the whole number is 1. To get the remainder,

$$\text{remainder} = 7 - (1 \cdot 4) = 7 - 4 = 3.$$

This gives

$$\frac{7}{4} = 1\frac{3}{4}.$$

Try this calculator technique with a more difficult example, say $\dfrac{386}{28}$

$$386 \div 28 = 13.7857$$

So now we know that the whole number is 13. To get the remainder,

$$\text{remainder} = 386 - (13 \cdot 28) = 386 - 364 = 22$$

This gives

$$\frac{386}{28} = 13\frac{22}{28} \ .$$

Terminology: When the remainder is zero, the numerator is said to be divisible by the denominator. For example,

$$\frac{8}{4} = \frac{8+0}{4} = \frac{8}{4} + \frac{0}{4} = 2\frac{0}{4} = 2$$

In the example above, 8 is divisible by 4 because when 8 is divided by 4 the remainder is zero.

Converting mixed numbers to improper fractions

Of course you are also expected to be able to convert a mixed number to an improper fraction.

$$2\frac{5}{8} = 2 + \frac{5}{8} = \frac{16}{8} + \frac{5}{8} = \frac{16+5}{8} = \frac{21}{8}$$

Another way to think of this is

$$2\frac{5}{8} = \frac{(2 \cdot 8) + 5}{8} = \frac{16+5}{8} = \frac{21}{8}$$

This is a better way to convert a mixed number to an improper fraction (it is faster and makes better use of the calculator). Take the whole number, multiply it by the denominator, then add the product to the numerator.

Word problems involving remainders

There is a class of problems on standardized tests that I call "remainder problems" because they are based on converting an improper fraction to a mixed number. Although there are several problems of this type in unit 6.3, it is worth while to take a look at them now as well. A good example is:

Sarah wanted to make jump ropes for herself and her friends, so she went to a store and purchased 100 feet of rope. If each jump rope is 7 feet long, how many jump ropes could Sarah make and how many feet of rope will she have left over?

Begin by dividing the 100 total feet of rope by each of the 7 foot lengths. Using the calculator, this gives

$$\frac{100}{7} = 14.2857$$

So now we know that there will be 14 jump ropes. To get the amount of rope left over:

$$\text{rope left over} = 100 - (14 \cdot 7) = 100 - 98 = 2$$

Mathematically what we just did was

$$\frac{100}{7} = 14\frac{2}{7}.$$

Two is the remainder, which in this problem represents the amount of rope left over.

Problems on mixed numbers and remainders (unit 2.3)

1. What is the remainder when 287 is divided by 5?
   (A)   2
   (B)   4
   (C)   20
   (D)   57
   (E)   285

2. What is the remainder when 980 is divided by 28?
   (A)   0
   (B)   1
   (C)   2
   (D)   3
   (E)   35

3. The mixed number $16\frac{7}{8}$ is equivalent to:
   (A)   23/8
   (B)   31/8
   (C)   112/8
   (D)   128/8
   (E)   135/8

4. The mixed number $-11\frac{6}{7}$ is equivalent to
   (A)   −66/7
   (B)   −71/7
   (C)   -77/7
   (D)   -83/7
   (E)   83/7

5. If 50 campers are assigned to red, white or blue teams, in that order, to which team is the last camper assigned?
   (A)   red
   (B)   white
   (C)   blue
   (D)   red or white
   (E)   red or blue

6. The night before Halloween, Kyle placed 1,000 pieces of candy into bags, with each bag containing 12 pieces of candy. If Kyle is allowed to eat any candy left over, how many pieces could he eat?
   (A)   0
   (B)   4
   (C)   8
   (D)   16
   (E)   83

7. Mary has $25 to make copies of her flyer. If each copy costs 8 cents, how many copies can she make?
   (A)   3
   (B)   4
   (C)   5
   (D)   31
   (E)   312

8. To get a discount, Bill has to buy at least $50 of newspapers. If each newspaper costs 75 cents, how many newspapers must he buy to get the discount?
   (A)   37
   (B)   38
   (C)   66
   (D)   67
   (E)   667

9. Emily's Mom sends her into the store with $10, telling her to buy as many apples as possible. If each apple costs 80 cents, how much change (in cents) will Mary receive?
   (A)   2
   (B)   4
   (C)   20
   (D)   40
   (E)   50

10. When an integer x is divided by 8, the remainder is 3. What cannot be the value of x?
   (A)   3
   (B)   19
   (C)   24
   (D)   67
   (E)   99

Solutions to problems on mixed numbers and remainders (unit 2.3)

1. (A)

Use your calculator to find $\dfrac{287}{5} = 57.4$. Now we know that the whole number is 57. Next find the remainder $287 - (57 \cdot 5) = 287 - 285 = 2$.

2. (A)

Use your calculator to find $\dfrac{980}{28} = 35.0$ The remainder is zero.

3. (E)

$16\dfrac{7}{8} = \dfrac{(16 \cdot 8) + 7}{8} = \dfrac{128 + 7}{8} = \dfrac{135}{8}$

4. (D)

$-11\dfrac{6}{7} = \dfrac{(-11 \cdot 7) - 6}{7} = \dfrac{-77 - 6}{7} = \dfrac{-83}{7}$.

5. (B)

One way to do this problem is to see that when the camper's number is divided by three: all campers assigned to the red team will have a remainder of 1, all campers assigned to the white team will have a remainder of 2, and all campers assigned to the blue team will have a remainder of 0. When 50 is divided by 3, the remainder is 2. Therefore the 50th camper will be assigned to the white team.

6. (B)

$\dfrac{1000}{12} = 83\dfrac{4}{12}$

Kyle makes 83 bags and eats 4 pieces of candy.

7. (E)

$\dfrac{25.0}{0.08} = 312\dfrac{4}{8}$

Mary can make 312 copies and she will have 4 cents left over.

8. (D)

$\dfrac{50}{0.75} = 66\dfrac{2}{3}$.

To exceed \$50, Bill must buy 67 newspapers.

9. (D)

$\dfrac{10.0}{0.8} = 12\dfrac{1}{2}$.

Emily will buy 12 apples and pay \$9.60. She will receive 40 cents change.

10. (C)

$\dfrac{24}{8} = 3\dfrac{0}{8}$. The remainder is zero.

Unit 2.4 Order of operations

Order of operations is a fairly important topic in its own right, but it is especially important for students who rely on the calculator. The calculator will always follow the order of operations no matter what, even if that is not what you intended it to do.

The order of operations is best remembered by the acronym, PEMDA:

P = parenthesis. Always perform operations inside parentheses first

E = exponents. Next, raise terms to their powers (exponents)

M = multiplication . Multiplication and division go together.

D = division. Multiply and divide terms, from left to right.

A = addition and subtraction. Last, add and subtract terms, left to right.

Like many things in math, examples are the best way to learn PEMDA. Start with

$$5 \cdot \left(8 + 2^2\right) \div 4 - 3^2$$

The first step is to simplify the expression within the parenthesis, giving us

$$5 \cdot \left(12\right) \div 4 - 3^2$$

Next, we clear the exponents

$$5 \cdot \left(12\right) \div 4 - 9$$

Next, we multiply and divide, left to right

$$60 \div 4 - 9$$
$$15 - 9$$

The last step is addition and subtraction, yielding an answer of 6.

PEMDA does not only apply to numbers, it also applies to variables. Consider

$$2x + \left(2x + x\right)^2 - 6x^2 \div 2x - x$$

First simplify expressions inside parentheses

$$2x + \left(3x\right)^2 - 6x^2 \div 2x - x$$

Next we clear out the exponents

$$2x + 9x^2 - 6x^2 \div 2x - x$$

Next divide

$$2x + 9x^2 - 3x - x$$

The last step is to add and subtract (combine like terms)

$$9x^2 - 2x$$

Calculator tips

On calculators PEMDA errors often involve negative numbers. First of all, please remember that there is a negative key (labeled as "(-)" on the calculator) and a subtraction key (labeled as "-" on the calculator). These keys are different. The negative key is used to set the sign of a number, whereas the subtraction key is used to subtract one number from another (either could be positive or negative).

Suppose you want to subtract negative 3 from 5

$$5 - -3 = 8.$$

The key sequence is 5, -, (-), 3, Enter. This gives you the correct result of 8. The calculator is following PEMDA. It first multiplies -1 by 3. Then it subtracts -3 from 5.

Suppose you wanted to calculate the square of negative 2. Rather than get the correct answer of positive 4 $(-2)^2 = (-2) \cdot (-2) = 4$, many students will get -4. This is because the calculator followed PEMDA but the student did not. If you enter (-), 2, x^2 the calculator will return -4

$$-2^2 = -2 \cdot 2 = -4.$$

This is due to PEMDA. The calculator deals with the <u>exponent first</u>, raising 2 to the second power, giving 4. Then it multiplies 4 by -1. To get the correct result of positive 4 you must use parentheses. The correct key sequence is (, (-), 2,), x^2 , enter.

$$(-2)^2 = 4.$$

Problems on order of operations (unit 2.4)

1. The expression $\dfrac{2-12\div 2}{-2^2}$ is equal to

    (A) $-\dfrac{5}{4}$

    (B) -1

    (C) -0.5

    (D) 1

    (E) $\dfrac{5}{4}$

2. The expression $5-2^2+3(1+2)^2$ is equal to:

    (A)   10

    (B)   28

    (C)   30

    (D)   36

    (E)   145

3. The expression $\dfrac{-4}{2}-(-2)^3+1$ is equal to:

    (A) -9

    (B) -7

    (C) -5

    (D) 7

    (E) 11

4. The expression $3^2+\dfrac{2^3+1}{(1-2)^3}$ is equal to:

    (A) -18

    (B) -9

    (C) 0

    (D) 9

    (E) 18

5. The expression $\dfrac{8+(6-3)^2}{2^2-3^2-12}$ is **closest** to:

    (A) -2

    (B) -1

    (C) 0

    (D) 1

    (E) 2

6. The expression $\dfrac{(3x)^3-x^3}{x^3}$ is equal to:

    (A) 0

    (B) 2

    (C) 8

    (D) 26

    (E) 52

7. The expression $-x^2-(-2x)^2-x+(-3)^3$ is equivalent to:

    (A) $3x^2-x+9$

    (B) $3x^2-x-27$

    (C) $7x^2-x+-27$

    (D) $-6x^2-27$

    (E) $-5x^2-x-27$

8. The expression $\dfrac{(x^2)^2+2x^2-6x\div 2x}{x\cdot x\cdot x-x\cdot x+x+x+x-3}$ is equivalent to:

    (A) $x+1$

    (B) x^2+3

    (C) $x-1$

    (D) x^2-3

    (E)   None of the above.

Solutions to problems on order of operations (unit 2.4)

1. (D)

$$\frac{2-12\div 2}{-2^2} = \frac{2-6}{-4} = \frac{-4}{-4} = 1$$

2. (B)

$$5-2^2 + 3(1+2)^2 = 5-2^2 + 3\cdot 3^2$$
$$= 5-4+3\cdot 9 = 5-4+27 = 28$$

3. (D)

$$\frac{-4}{2} - (-2)^3 + 1 = -2 - (-8) + 1 = -2 + 8 + 1 = 7$$

4. (C)

$$3^2 + \frac{2^3 + 1}{(1-2)^3} = 9 + \frac{8+1}{-1} = 9 - 9 = 0$$

5. (B)

$$\frac{8 + (6-3)^2}{2^2 - 3^2 - 12} = \frac{8 + 3^2}{2^2 - 3^2 - 12} = \frac{8+9}{4-9-12} = \frac{17}{-17} = -1$$

6. (D) $\dfrac{(3x)^3 - x^3}{x^3} = \dfrac{27x^3 - x^3}{x^3} = \dfrac{26x^3}{x^3} = 26$

7. (E)

$$-x^2 - (-2x)^2 - x + (-3)^3 = -x^2 - 4x^2 - x - 27$$
$$= -5x^2 - x - 27$$

8. (A)

$$\frac{\left(x^2\right)^2 + 2x^2 - 6x \div 2x}{x\cdot x\cdot x - x\cdot x + x + x + x - 3} = \frac{x^4 + 2x^2 - 3}{x^3 - x^2 + 3x - 3}$$
$$= \frac{\left(x^2 + 3\right)\left(x^2 - 1\right)}{x^2\left(x - 1\right) + 3\left(x - 1\right)}$$
$$= \frac{\left(x^2 + 3\right)(x+1)(x-1)}{\left(x^2 + 3\right)(x-1)}$$
$$= x + 1$$

Unit 2.5 Percentages

Percentages are a favorite topic for standardized tests, and they can come in several forms. How many ways can we write five percent? Three ways are possible:

$$5\%, .05 \text{ and } \frac{5}{100}.$$

The last way, as a fraction, is the most important to remember.

How many ways can we write x percent? Two ways are possible:

$$x\% \text{ and } \frac{x}{100}.$$

The last way, as a fraction, is the only useful way to write x percent. It is critical to remember that because in word problems the phrase "what percent" is substituted by $\frac{x}{100}$ when we set up an equation. Consider:

What percent of 86 is 18?

This is not difficult if you substitute correctly. The equation becomes:

$$\left(\frac{x}{100}\right)86 = 18, \ 86x = 1800, \ x = \frac{1800}{86}, \ x = 20.93$$

Consider another simple word problem:

28 is 16 percent of what number?

When written as an equation we have:

$$28 = \left(\frac{16}{100}\right)x, \ 2800 = 16x, \ \frac{2800}{16} = x, \ 175 = x$$

It is faster (and better) to approach this problem directly, using decimals:

$$28 = 0.16x, \ \frac{28}{.16} = x, \ 175 = x$$

Lastly we consider an even simpler word problem

What is 5% of 185?

If we were to set up an equation, we would have:

$$\left(\frac{5}{100}\right)185 = \frac{925}{100} = 9.25$$

It is faster (and better) to solve using decimal notation:

$$(.05)185 = 9.25$$

With practice, you will gain insight into when it is better to use the decimal or fraction representations of percentage amounts. You should begin by using the fraction and then start using the decimal amounts to speed up calculations.

Percentage Increase

Just as there are two forms to remember for percentages (as decimals and fractions), there are two forms to remember for percentage increase, and both forms are useful. First, we consider:

$$\text{percentage increase} = \left(\frac{new - old}{old}\right)100$$

If a store raises its price for a shirt from \$80 to \$90, the percentage increase is

$$\left(\frac{90-80}{80}\right)100 = \left(\frac{10}{80}\right)100 = \frac{1000}{80} = 12.5$$

Another useful form of percentage increase is that to:

$$\text{inflate y by x\%} = \left(1 + \frac{x}{100}\right)y$$

If a store decides to raise all of its prices by 5%, then a shirt that originally costs \$80 will now cost

$$\left(1 + \frac{5}{100}\right)80 = (1.05)80 = 84 \ .$$

Percentage Decrease

The equations for percentage decrease are similar to the ones above for percentage increase. The first is:

$$\text{percentage decrease} = \left(\frac{old - new}{old}\right)100$$

If a store marks its price for a shirt from \$90 to \$80, the percentage decrease is

$$\left(\frac{90-80}{90}\right)100 = \left(\frac{10}{90}\right)100 = \frac{1000}{90} = 11.1$$

The other useful form of percentage is that to:

$$\text{deflate y by x\%} = \left(1 - \frac{x}{100}\right)y$$

If a store decides to reduce all of its prices by 5%, then a shirt that originally costs \$80 will now cost

$$\left(1 - \frac{5}{100}\right)80 = (0.95)80 = 76 \ .$$

Problems on percentages (unit 2.5)

1. What is 28% of 600?

(A)   0.47

(B)   21.42

(C)   168

(D)   2,142

(E)   16,800

2. 15 is what percent of 500?

(A)   0.03

(B)   3

(C)   33

(D)   75

(E)   7,500

3. 65 is 20 percent of what number?

(A)   0.31

(B)   3.25

(C)   13

(D)   325

(E)   1,300

4. Suppose that 100 years ago, the price of a pound of sugar was 25 cents. It is now $3. How much has the price changed?

(A)   0.9%

(B)   11%

(C)   91.6%

(D)   1100%

(E)   1199%

5 Diana chose a new pair of shoes marked $120, and the store is having a sale of 20% off everything. How much will she pay for the shoes (ignoring sales tax)?

(A) $24

(B) $96

(C) $114

(D) $144

(E) $240

6. Two years ago the average rain fall was 30 inches. If it has increased by 15% per year since then, what is the average annual rainfall now?

(A)   30.0

(B)   34.5

(C)   39.0

(D)   39.7

(E)   67.5

7. A store clerk was instructed to mark everything up by 15%. Instead he marked everything down by 15%. If a pair of shoes was originally marked $80, what is difference between the correct price and the incorrect price marked by the clerk?

(A) $8

(B) $12

(C) $24

(D) $68

(E) $92

8. Goeff bought a hat with a price tag of $15.50. The sales clerk charged $11 for the hat, saying that there is a manager's sale today. What percentage off was the sale price?

(A)   13.25%

(B)   26%

(C)   29%

(D)   36%

(E)   41%

9. Your have traveled to a state where they charge 10% sales tax. A store window announces that everything is 10% off today. If you go into that store to buy a pair of shoes marked $65, how much will you pay, including sales tax?

(A) $7.15

(B) $58.50

(C) $64.35

(D) $65.00

(E) $71.50

Solutions to problems on percentages (unit 2.5)

1. (C)

$$\left(\frac{28}{100}\right)600 = \frac{16,800}{100} = 168$$

2. (B)

$$15 = \left(\frac{x}{100}\right)500, \ 1500 = 500x, \ 3 = x;$$

or use ratios: $\dfrac{15}{500} = \dfrac{x}{100}$

3. (D)

$$65 = \left(\frac{20}{100}\right)x, \ 6500 = 20x, \ \frac{6500}{20} = x, \ 325 = x$$

4. (D)

$$\left(\frac{3.0 - 0.25}{0.25}\right)100 = \left(\frac{2.75}{0.25}\right)100 = (11)100 = 1,100\%$$

5. (B)

$$\left(1 - \frac{20}{100}\right)120 = (.80)120 = 96$$

You could use the other formula for percent decrease, but it is more work.

$$\left(\frac{120 - x}{120}\right)100 = 20$$

6. (D)

$$\left(1 + \frac{15}{100}\right)30 = (1.15)30 = 34.5 \text{ after one year.}$$

$$\left(1 + \frac{15}{100}\right)34.5 = (1.15)34.5 = 39.7 \text{ after two years.}$$

7. (C)

The price should have been $80\left(1 + \dfrac{15}{100}\right) = 92$.

Instead the customer was charged

$$80\left(1 - \frac{15}{100}\right) = 80(.85) = 68.$$

So the store lost 92-68=24 dollars.

8. (C)

$$\left(\frac{15.5 - 11}{15.5}\right)100 = \left(\frac{4.5}{15.5}\right)100 = 29\% \text{ off}$$

9. (C)

First calculate the price with the discount

$$\left(1 - \frac{10}{100}\right)65 = (.90)65 = 58.5. \text{ Next calculate}$$

how much you pay including sales tax

$$58.5\left(1 + \frac{10}{100}\right) = 58.5(1.1) = 64.35. \text{ Notice that}$$

you will not pay the original price of $65, but something slightly less.

Unit 2.6 Averages

Like many so-called simple things, the standardized tests can take something simple like an average and stand it on its head.

We all know that

$$average = \frac{sum}{count}.$$

So, the average of 5 and 8 is $\frac{5+8}{2} = \frac{13}{2} = 6.5$ and the average of x and y would be $\frac{x+y}{2}$.

In general, when you see the word "average" on a standardized test, you should think about the sum. Many word problems involving averages are really about sums, and the following form of the sum is useful to keep in mind.

$$sum = (average) \cdot (count)$$

Consider this word problem:

> Two people are on an elevator, and their average weight is 200 pounds. The elevator has a capacity of 500 pounds. The elevator comes to stop and a third person wants to get in. What is the maximum amount that the third person can weigh?

The solution comes from thinking in terms of sums. If the two people on the elevator have an average weight of 200 pounds, the sum of their weights must be 400 pounds. They could each weigh 200 pounds, or one could weigh 50 pounds while the other weighs 350 pounds. It does not matter what either weighs. What matters is that together they weigh 400 pounds. As the elevator has a capacity of 500 pounds, the third would-be passenger cannot weigh more than 100 pounds.

Averages and sequences

There is more on sequences in unit 5.1, but arithmetic sequences have special properties related to the average. An arithmetic sequence is one which is evenly spread, where the difference between any two adjacent terms is always the same. This is called the common difference. An example of an arithmetic sequence is

$$8, 12, 16, 20, 24.$$

The common difference is 4.

It turns out that the median of an arithmetic sequence is the average. In the example above, the median is 16 and so is the average. You can check by adding up the numbers on your calculator and dividing by 5. If the average is 16 then the sum must be $16 \cdot 5 = 90$. For an arithmetic sequence, we can find the average and sum quickly, even without a calculator.

A special case of the arithmetic sequence (and a favorite for standardized tests) is where the median is zero

$$-15, -10, -5, 0, 5, 10, 15$$

The average is zero and the sum is zero. Keep a sharp lookout for these.

Problems on averages (unit 2.6)

1. If the average of x, y and z is 15, what is their sum?
   (A)   5
   (B)   15
   (C)   30
   (D)   45
   (E)   60

2. What is the sum of 5, 7, 9, 11, 13, 15 and 17 (do not use a calculator)?

   (A)   5
   (B)   7
   (C)   9
   (D)   11
   (E)   77

3. If the sum of nine consecutive integers is zero, what is the largest integer?
   (A)   0
   (B)   1
   (C)   2
   (D)   3
   (E)   4

4. Twelve students took a test and their average score was 85. However, Bob was sick that day. After Bob took the test, the average score dropped to 82. What was Bob's score on the test?
   (A)   3
   (B)   36
   (C)   46
   (D)   56
   (E)   Cannot be determined

5. If the average if x and y is 12 and z=6, what is the average of x, y, and z?.
   (A)   6
   (B)   10
   (C)   24
   (D)   30
   (E)   Cannot be determined

6. If the average of 120 consecutive integers is -20, what is their sum?
   (A)   -0.17
   (B)   -6
   (C)   -24
   (D)   -2400
   (E)   Cannot be determined

7. An expression for the average of x, 2x, 3y is
   (A) $3x + 3y$

   (B) $\dfrac{x+y}{2}$

   (C) $\dfrac{x+y}{3}$

   (D) $x + y$

   (E) $\dfrac{3}{2}(x+y)$

8. The average of 11 consecutive odd integers is -3. What is the smallest integer?
   (A)   -15
   (B)   -13
   (C)   -11
   (D)   -8
   (E)   2

Solutions to problems on averages (unit 2.6)

1. (D)

$$\frac{x+y+z}{3}=15, \quad x+y+z=45$$

2. (E)

The median is 11 so the average must be 11 because this is an arithmetic sequence with a common difference of 2. There are 7 numbers so their sum must be $7 \cdot 11 = 77$.

3. (E)

Consecutive integers form an arithmetic sequence with a common difference of 1. If their sum is zero then the median must be zero. So the sequence must be -4, -3, -2, -1, 0, 1, 2, 3, 4. The largest integer is 4.

4. (C)

Before Bob took the test, the class sum was $85 \cdot 12 = 1020$. After Bob took the test, the class sum was $82 \cdot 13 = 1066$. Therefore Bob's score on the test was $1066 - 1020 = 46$. Bob didn't do very well.

5. (B)

If the average of x and y is 12, then the sum of x and y is 24. The sum of x, y and z is 24+6=30. The average of x, y, and z is 30/3 = 10.

6. (D)

If the average of 120 numbers is -20 then their sum must be $120(-20) = -2400$. It does not matter that these are consecutive integers.

7. (D).

The average is
$$\frac{x+2x+3y}{3} = \frac{3x+3y}{3} = x+y$$

8. (B)

This constitutes an arithmetic sequence with a common difference of 2, so the average is the median is -3. Therefore there are five numbers to either side of -3. The sequence is -13, -11, -9, -7, -5, -3, -1, 1, 3, 5, 7. The smallest term -13.

3. ALGEBRA 1

Congratulations, you have now graduated from middle school! Seriously though, that early math can be tricky. Now you are ready for some algebra.

Unit 3.1 Solving equations and inequalities

Algebra is mostly about solving equations. Sometimes this is called balancing equations rather than solving equations, because whatever you do to one side of an equation must also be done to the other side of the equation.

The first technique to be learned is combining like terms. This means moving terms from one side of the equation to the other so that they can be combined. Start with a simple equation such as

$x - 3 = 12$ original equation

$x - 3 + 3 = 12 + 3$ add 3 to both sides (combine constant terms)

$x = 15$ simplify

A somewhat more complicated example of combining like terms:

$x + 19 = 4x - 2$ original equation

$x + 19 - x = 4x - 2 - x$ subtract x from both sides (combine x-terms)

$19 = 3x - 2$ simplify by combining like terms

$19 + 2 = 3x - 2 + 2$ add 2 to both sides (combine constant terms)

$21 = 3x$ simplify by combining like terms

$\dfrac{21}{3} = \dfrac{3x}{3}$ divide both sides by 3

$7 = x$ simplify

The other technique to be learned is opposites. This was used in the example just above. We use opposites to isolate the variable to be solved. In the equation $21 = 3x$ the variable x is being multiplied by 3. The opposite of multiplication is division, so we isolate the variable by dividing both sides of the equation by 3, yielding $7 = x$. In the equation $\dfrac{x}{5} = 3$ the variable is being divided by 5. The opposite of division is multiplication, so we isolate the variable by multiplying both sides of the equation by 5, yielding $x = 15$.

Inequalities are equations where the equals sign = is not used. Instead any of the following signs may be used:

      $<$ less than

      $\leq$ less than or equal to

      $>$ greater than

      $\geq$ greater than or equal to

The same two techniques are used to solve inequalities: combining like terms and opposites. The only difference is that when opposites involve a negative quantity, you must remember to change the direction of the inequality. In other words, multiplying or dividing an inequality by a negative number causes the direction of the inequality to change.

Consider the inequality:

$$-2(x-8) \leq 4 \qquad \text{original equation}$$

$$-2x + 16 \leq 4 \qquad \text{distribute the -2}$$

$$-2x \leq -12 \qquad \text{combine like terms (add 16 to both sides)}$$

$$x \geq 6 \qquad \text{isolate the variable (divide both sides by -2, change direction)}$$

The last step uses opposites to isolate the variable. Because both sides of the equation are divided by negative two, the direction of the inequality is changed.

Sometimes you are required to write the solution using interval notation or identify the graph of the solution on a number line. The "trick" here is to use a closed endpoint if the $\leq$ or $\geq$ sign is present. Use an open endpoint if the $<$ or $>$ sign is present. The examples below are helpful.

Inequality	Interval	Number Line
$x \geq 1$	$[1, \infty)$	●———————→ -4 -3 -2 -1 0 1 2 3 4
$x < -2$	$(-\infty, -2)$	←———○ -4 -3 -2 -1 0 1 2 3 4
$-3 \leq x < 3$	$[-3, 3)$	●————————○ -4 -3 -2 -1 0 1 2 3 4

Some students find it helpful to remember their brackets this way: when the bracket is square, it reaches out and includes the endpoint; whereas a rounded bracket pushes back and excludes the endpoint.

When using interval notation and positive infinity or negative infinity is an endpoint, the interval is always open because infinity can never be reached. A bracket can never get its arms around infinity or negative infinity.

Problems on solving equations and inequalities (unit 3.1)

1. The solution to the equation

$4(x+2)=3(5-x)$ is:

    (A) $x=7/15$

    (B) $x=1$

    (C) $x=13/7$

    (D) $x=7$

    (E) $x=15$

2. The solution to the equation

$\dfrac{3(x-6)}{5}=x$ is:

    (A) $x=-9$

    (B) $x=-3$

    (C) $x=0$

    (D) $x=9/4$

    (E) $x=3$

3. The solution to the equation

$\dfrac{4x}{3}=2(x+9)$ is:

    (A) $x=-54$

    (B) $x=-27$

    (C) $x=-6.75$

    (D) $x=-4.5$

    (E) $x=0$

4. The solution to the inequality

$2x+7<-11$ is

    (A) $x>2$

    (B) $x>9$

    (C) $x<-2$

    (D) $x<-9$

    (E)   None of the above

5. John must save at least \$25.50 to buy a hat. If he saves 80 cents per day, how many days will it take him to save enough?

    (A)   3

    (B)   4

    (C)   30

    (D)   31

    (E)   32

6. The solution to the inequality $5-\dfrac{x}{2}\le 4$ is:

    (A) $[2,\infty]$

    (B) $(2,\infty)$

    (C) $[2,\infty)$

    (D) $(-\infty,1]$

    (E) $[-\infty,1]$

7. The solution to the inequality $2\le 2x+1<3$ is:

    (A) $(-0.5,1]$

    (B) $[-0.5,1)$

    (C) $[-0.5,1]$

    (D) $(0.5,1]$

    (E) $[0.5,1)$

8. The solution to the inequality $6x-3<-9$ is

    (A) $x<1$

    (B) $x>1$

    (C) $x<-1$

    (D) $x>-1$

    (E) $x<-2$

9. Mary and Sarah want to get on a ride at the amusement park that can hold no more than 100 pounds. If Mary weighs 35 pounds, what is the most that Sarah can weigh?

    (A)   64 pounds

    (B)   64.99999 pounds

    (C)   65 pounds

    (D)   65.1 pounds

    (E)   66 pounds

Solutions to problems on solving equations and inequalities (unit 3.1)

1. (B).

$$4(x+2)=3(5-x)$$
$$4x+8=15-3x$$
$$7x=15-8=7$$
$$x=1$$

2. (A)

$$\frac{3(x-6)}{5}=x$$
$$3(x-6)=5x$$
$$3x-18=5x$$
$$-18=2x$$
$$-9=x$$

3. (B)

$$\frac{4x}{3}=2(x+9)$$
$$4x=6(x+9)$$
$$4x=6x+54$$
$$-2x=54$$
$$x=-27$$

4. (D)

$$2x+7<-11$$
$$2x<-18$$
$$x<-9$$

5. (E)

$$.80d \geq 25.5$$
$$d \geq 31.875$$

John will not have enough money until the 32nd day.

6. (C)

$$5-\frac{x}{2}\leq 4$$
$$-\frac{x}{2}\leq -1$$
$$x \geq 2 \text{ or } [2,\infty)$$

7. (E)

$$2 \leq 2x+1 \text{ and } 2x+1<3$$
$$1 \leq 2x \text{ and } 2x<2$$
$$0.5 \leq x \text{ and } x<1, \text{ the interval is } [0.5,1)$$

8. (C)

$$6x-3<-9$$
$$6x<-6$$
$$x<-1$$

9. (C)

$$x+35 \leq 100$$
$$x \leq 65$$

Unit 3.2 Square roots (radicals)

Remember that the square root symbol $\sqrt{}$ means the positive square root. There is an invisible + sign in front of the radical. Although it is invisible, it is very important. For example, $\sqrt{9}$ is positive 3, not negative 3. In words, this would be "the positive square root of nine is positive three, not negative three."

When we mean to indicate the negative square root, we write $-\sqrt{9} = -3$. In words, this would be "the negative square root of nine is negative three." When we mean to indicate the positive **and** negative square roots, we write $\pm\sqrt{9} = \pm 3$. In words, this would be "the positive and negative square roots of nine are positive and negative three."

Positive and negative square roots are used mostly for solving equations, such as

$$x^2 = 64$$
$$x = \pm\sqrt{64}$$
$$x = \pm 8$$

In words, this would be "the solutions to the equation $x^2 = 64$ are $x = 8$ and $x = -8$". Remember that the $\pm$ symbol means positive AND negative. So the symbol $\pm\sqrt{}$ means the positive and negative square root, while the symbol ± 8 means positive 8 AND negative 8.

Check these:

$$\sqrt{25} = 5 \text{ only}$$
$$-\sqrt{25} = -5 \text{ only}$$
$$\pm\sqrt{25} = 5 \text{ and -5 (you could also write } \pm 5)$$

Remember:

$\pm$ means positive AND negative

$\sqrt{}$ means positive square root

General radicals

There is more material on radicals in unit 5.3. But for now, just be aware that the positive square root symbol has another invisible element: there is an invisible two. The symbol $\sqrt{}$ is actually $+\sqrt[2]{}$. Please note the invisible positive sign and the invisible two.

In general, the positive nth root is symbolized by $\sqrt[n]{}$, where n is a positive integer.

For example to find the cube root of 8, we write $\sqrt[3]{8} = x$ if $x \cdot x \cdot x = 8$.

Nth roots of numbers can be found easily on the calculator using fractional exponents because of the identity:

$$\sqrt[n]{x} = x^{\frac{1}{n}}$$

To find the 5th root of 7776, simply enter 7776, ^, (, 1, ÷, 5 ,) into your calculator and then press enter. Your display should look like:

In other words $6^5 = 7776$.

Laws of radicals

There are two laws of radicals to be covered here. The rest are covered in Chapter 5. These laws are typically used to simplify radicals. They are:

$$\sqrt[n]{xy} = \sqrt[n]{x} \cdot \sqrt[n]{y}$$

$$\sqrt[n]{\frac{x}{y}} = \frac{\sqrt[n]{x}}{\sqrt[n]{y}}$$

Some examples of the multiplication law are:

$$\sqrt{50} = \sqrt{50 \cdot 2} = \sqrt{25} \cdot \sqrt{2} = 5\sqrt{2}$$

$$\sqrt[3]{192} = \sqrt[3]{64 \cdot 3} = \sqrt[3]{64} \cdot \sqrt[3]{3} = 4\sqrt[3]{3}$$

An example of the division law:

$$\sqrt{\frac{5}{12}} = \frac{\sqrt{5}}{\sqrt{12}} = \frac{\sqrt{5}}{\sqrt{4 \cdot 3}} = \frac{\sqrt{5}}{\sqrt{4} \cdot \sqrt{3}} = \frac{\sqrt{5}}{2\sqrt{3}}$$

In general, we do not like to have a radical in a denominator, so this should be further simplified as follows:

$$\frac{\sqrt{5}}{2\sqrt{3}} = \left(\frac{\sqrt{5}}{2\sqrt{3}}\right)\left(\frac{\sqrt{3}}{\sqrt{3}}\right) = \frac{\sqrt{15}}{2 \cdot 3} = \frac{\sqrt{15}}{6}$$

Problems on radicals (unit 3.2)

1. $\sqrt{49}$ is equivalent to:
   - (A)   7 only
   - (B)   -7 only
   - (C)   -7 or 7
   - (D)   -7 and 7
   - (E)   all of the above

2. $\sqrt[3]{-27}$ is equivalent to:
   - (A)   3 only
   - (B)   -3 only
   - (C)   -3 or 3
   - (D)   -3 and 3
   - (E)   all of the above

3. Which of the following is the solution to the equation $x^2 - 5 = 20$?
   - (A)   5 only
   - (B)   -5 only
   - (C)   -5 and 5
   - (D) $\sqrt{15}$
   - (E) $\pm\sqrt{15}$

4. $\sqrt{72}$ is equivalent to:
   - I. $6\sqrt{2}$
   - II. $3\sqrt{8}$
   - III. $2\sqrt{18}$
   - (A)   I only
   - (B)   II only
   - (C)   III only
   - (D)   I and III only
   - (E)   I, II and III

5. $\sqrt[3]{-384}$ is equivalent to:
   - (A) $8\sqrt{6}$
   - (B) $-8\sqrt{6}$
   - (C) $4\sqrt{24}$
   - (D) $4\sqrt[3]{6}$
   - (E) $-4\sqrt[3]{6}$

6. The solution to the equation $8 + \sqrt{2x} = 16$ is:
   - (A) $x = 4$
   - (B) $x = 8$
   - (C) $x = 16$
   - (D) $x = 32$
   - (E) $x = 64$

7. The solution to the equation $1 - \sqrt{x} = \dfrac{-7}{2}$ is:
   - (A) $x = 1.5$
   - (B) $x = 2$
   - (C) $x = 4$
   - (D) $x = 16$
   - (E) $x = 20.25$

8. 497 is ten less than three times the square of a number. The value of that number is:
   - I. -13
   - II. 13
   - III. 12.74
   - (A)   I only
   - (B)   II only
   - (C)   III only
   - (D)   I and II only
   - (E)   None of the above

Solutions to practice problems on radicals (unit 3.2)

1. (A)

$$\sqrt{49} = +7$$

2. (B)

$$\sqrt[3]{-27} = -3 \text{ because } (-3)(-3)(-3) = -27$$

3. (C)

$$x^2 - 5 = 20$$
$$x^2 = 25$$
$$x = \pm\sqrt{25} = \pm 5$$

4. (E)

Start by dividing perfect squares into 72

$$\sqrt{72} = \sqrt{36 \cdot 2} = \sqrt{36} \cdot \sqrt{2} = 6\sqrt{2}$$
$$\sqrt{72} = \sqrt{9 \cdot 8} = \sqrt{9} \cdot \sqrt{8} = 3\sqrt{8}$$
$$\sqrt{72} = \sqrt{4 \cdot 18} = \sqrt{4} \cdot \sqrt{18} = 2\sqrt{18}$$

5. (E)

Start by dividing perfect cubes into 384

$$\sqrt[3]{-384} = \sqrt[3]{-64 \cdot 6} = \sqrt[3]{-64} \cdot \sqrt[3]{6} = -4\sqrt[3]{6}$$

6. (D)

$$8 + \sqrt{2x} = 16, \quad \sqrt{2x} = 8, \quad \left(\sqrt{2x}\right)^2 = 8^2, \quad 2x = 64, \quad x = 32$$

7. (E)

$$1 - \sqrt{x} = \frac{-7}{2}, \quad \sqrt{x} = \frac{9}{2}, \quad x = \frac{81}{4} = 20.25$$

8. (D)

$$3x^2 - 10 = 497$$
$$3x^2 = 507$$
$$x^2 = 169$$
$$x = \pm 13$$

Unit 3.3 Solving pairs of equations

Unit 3.1 dealt with solving one equation with one variable. But how do you handle two equations with two variables? There are two techniques to know: substitution and linear combination.

Substitution

Substitution is a very simple way to do the job. When given a pair of equations, re-write one of the equations to isolate one of the variables. Then substitute that variable into the second equation. For example, consider the pair of equations:

$$x + 2y = 10$$
$$2x + 3y = 500$$

Substitution for this pair of equations is simple because the first equation can be re-written as $x = 10 - 2y$. Next replace x with $10 - 2y$ in the second equation:

$$2x + 3y = 500$$
$$2(10 - 2y) + 3y = 500$$
$$20 - 4y + 3y = 500$$
$$y = -480$$

Finish the problem by finding the value of x:

$$x = 10 - 2y = 10 - 2(-480) = 10 + 960 = 970$$

Elimination

The other method of solving a pair of equations is elimination. First, one or both equations are multiplied by constants. Next, the equations are added to one another or subtracted from one another. In our example above, we could have first multiplied the top equation by two, yielding:

$$2x + 4y = 20$$
$$2x + 3y = 500$$

Next we can subtract the second equation from the first equation, giving

$$2x - 2x + 4y - 3y = 20 - 500$$
$$y = -480$$

In general, elimination is faster than substitution, but more prone to error. Sometimes it is necessary to multiply both equations by constants, as you can see in the next example.

$$3x + 2y = 255$$
$$2x + 5y = 500$$

Multiply the top equation by two and the bottom equation by three. This yields

$$6x + 4y = 510$$
$$6x + 15y = 1500$$

Subtracting the bottom equation gives

$$4y - 15y = 510 - 1500$$
$$-11y = -990, \quad y = 90$$

Substituting 90 for y in the first equation gives

$$3x + 2 \cdot 90 = 255, \quad 3x = 75, \quad x = 25 .$$

Calculator Tip: Pairs of equations may also be solved on the calculator. First, on paper, you must put each equation into y= format, isolating the y variable. Then enter each equation into the equation editor. Now press the graph key. Each equation is represented by a line on the graph. The intersection of the lines represents the solution. Once the two lines appear on the graph, the point of intersection may be found by using 2nd-calc-intersect.

Triples

Sometimes you will see a system of three equations with three variables. Do not let this bother you. Simply combine equations to create a new system of two equations with two variables. Then solve the reduced pair. For example

$$x + y - z = 40$$
$$2x - y + z = 80$$
$$2x + y + z = 100$$

Combining the top two equations gives $3x = 120$, whereas combining the bottom two equations gives $-2y = -20$. These are solved easily.

Important Tip: When a problem on the test contains two or more equations, always stack the equations vertically so that the variables are aligned. That way, you can see what technique will be best.

Problems on solving pairs of equations (unit 3.3)

1. If $x + y = 3$ and $3x + 5y = 13$, then $x - y$ is equal to:

    (A)   -2

    (B)   -1

    (C)   1

    (D)   2

    (E)   3

2. Tickets to a show cost \$3 for students and \$7 for adults. If 250 tickets were sold and ticket sales totaled \$1102, how many adult tickets were sold?

    (A)   35

    (B)   88

    (C)   111

    (D)   162

    (E)   352

3. If $3x + 5y = 13$ and $6x + 7y = 20$, find the value of $\dfrac{y}{x}$.

    (A)   0.5

    (B)   1

    (C)   1.5

    (D)   2

    (E)   2.5

4. If $5x + 2y = 8$ and $3x - 7y = 13$, find the value of $x - y$.

    (A)   -2

    (B)   -1

    (C)   1

    (D)   2

    (E)   3

5. Solve for x:
$$x + 2y + z = 3$$
$$2x - y - z = -5$$
$$3x - y + z = -2$$

    (A)   -2

    (B)   -1

    (C)   0

    (D)   1

    (E)   2

6. Find the average of x, y, and z:
$$x + 2y - z = -8$$
$$2x + y + 2z = 19$$
$$2x + 2y + 4z = 34$$

    (A)   0

    (B)   1

    (C)   3

    (D)   9

    (E)   45

7. Joanna saves nickels and quarters in a jar. If she saved five times as many nickels as quarters, and her coin collection is worth \$390, how many quarters has she saved?

    (A)   195

    (B)   300

    (C)   780

    (D)   1500

    (E)   7500

8. Lizzy's cell phone plan charges 5 cents per message for in-network text messages and 10 cents per message for out-of-network text messages. If she sent a total of 1500 messages and her bill was \$80, how many out-of-network messages did she send?

    (A)   50

    (B)   55

    (C)   60

    (D)   100

    (E)   1400

Solutions to problems on solving pairs of equations (unit 3.3)

1 (B)

Use substitution:

$$3x + 5y = 13$$

$$3(3-y) + 5y = 13$$

$$9 - 3y + 5y = 13, \quad y = 2, \quad x = 3 - 2 = 1$$

$$x - y = 1 - 2 = -1$$

2. (B)

Use substitution:

$$S + A = 250$$

$$3S + 7A = 1102$$

$$3(250 - A) + 7A = 1102$$

$$750 + 4A = 1102, \quad A = 88$$

3. (D)

Multiply the top equation by 2, then subtract:

$$3x + 5y = 13$$

$$6x + 7y = 20$$

$$10y - 7y = 26 - 20, \quad 3y = 6, \quad y = 2$$

$$3x + 5(2) = 13, \quad x = 1, \quad y/x = 2/1 = 2$$

4. (E)

Multiply top by 3, bottom by 5, then subtract:

$$5x + 2y = 8$$

$$3x - 7y = 13$$

$$15x + 6y = 24$$

$$15x - 35y = 65$$

$$6y + 35y = 24 - 65, \quad 41y = -41, \quad y = -1, \quad x = 2$$

$$x - y = 2 - -1 = 3$$

5. (B)

Reduce to two equations by eliminating z, then solve. First add the top two then add the bottom two.

$$x + 2y + z = 3$$

$$2x - y - z = -5$$

$$3x - y + z = -2$$

$$3x + y = -2$$

$$5x - 2y = -7$$

Next multiply the top equation by 2, then add:

$$6x + 5x = -4 - 7, \quad 11x = -11, \quad x = -1$$

6. (C)

Add all of the equations together to find the sum. Then divide by three to find the average.

$$5x + 5y + 5z = -8 + 19 + 34$$

$$5x + 5y + 5z = 45$$

$$x + y + z = 9$$

$$\frac{x + y + z}{3} = \frac{9}{3} = 3$$

7. (C)

$$5Q = N$$

$$.25Q + .05N = 390$$

$$.25Q + .05(5Q) = 390$$

$$.50Q = 390, \quad Q = 780$$

8. (D)

$$5x + 10y = 8000$$

$$x + y = 1500, \quad x = 1500 - y$$

$$5(1500 - y) + 10y = 8000$$

$$7500 - 5y + 10y = 8000$$

$$5y = 500, \quad y = 100$$

Unit 3.4 Distributing and common factors

Distribution involves multiplying a quantity across parentheses. For example:

$$3(2x-3y) = (3 \cdot 2x) - (3 \cdot 3y) = 6x - 9y$$

Distribution problems can involve a combination of constants and variables. For example:

$$6xy(2x+3y) = (6xy \cdot 2x) + (6xy \cdot 3y) = 12x^2y + 18xy^2$$

Factoring

Once the art of distributing is mastered, you are ready for factoring. Factoring is the opposite of distributing.

Suppose we want to factor the expression $6x - 9y$. Look for the factors that each term has in common:

$$6x = 2 \cdot 3 \cdot x \qquad 9y = 3 \cdot 3 \cdot y$$

The only factor in common to $6x$ and $9y$ is 3. So now we take out the common factor:

$$6x - 9y = 3(2x - 3y).$$

Consider the expression $12x^2y + 18xy^2z$. Break each term into its factors and look for the factors that are common to both:

$$12x^2y = 2^2 \cdot 3 \cdot x \cdot x \cdot y \qquad 18xy^2z = 2 \cdot 3^2 \cdot x \cdot y \cdot y \cdot z$$

The common factors are $2 \cdot 3 \cdot x \cdot y$. We take these common factors out of the terms and get

$$12x^2y + 18xy^2z = 2 \cdot 3 \cdot xy(2x + 3yz) = 6xy(2x + 3yz).$$

Greatest Common Factor (GCF)

In the example above, the constant terms we were dealing with were 12 and 18. We could have factored out 2, leaving 6 and 9. But 6 and 9 have a common factor of 3. We would then have to factor out the 3. It is far better to take out the greatest common factor (GCF) in one step.

Suppose we want to find the greatest common factor for 72 and 756. The greatest common factor can be found by following these steps:

1. Use a factor tree to find the prime factorization for each term (if you forgot how, see unit 2.1).
2. Make a list of the prime factors that appear in every term.
3. Raise each of these factors to the smallest power that is used.

For the example of 72 and 756, we have the following:

1. $72 = 2^3 \cdot 3^2$ and $756 = 2^2 \cdot 3^3 \cdot 7$.
2. The common primes are 2 and 3.
3. The smallest power of 2 is 2, and the smallest power of 3 is 2.
4. The GCF is $2^2 \cdot 3^2 = 4 \cdot 9 = 36$.

<u>Calculator Tip:</u> This technique works for finding the GCF for any number of terms. However, if you only need to find the GCF for two terms the "best way" is to use your calculator. To find the GCF of 72 and 756, simply enter the following sequence: *math-num-gcd-72-,-756-)-enter.*

Common Factors in Division

So far we have looked at how to use common factors in multiplication. Using the constants from the GCF example above, we have

$$72xy - 756x = 36x(2y - 21).$$

But what if we were given a division problem? The process is much the same:

$$\frac{756x}{72xy} = \left(\frac{36x}{36x}\right)\left(\frac{21}{2y}\right) = \frac{21}{2y}$$

Whether multiplying or dividing, we pull out the GCF, which in this case is $36x$.

When doing division, many students find it easier to break the fraction into stacks, sort of how we stacked pairs of equations in the previous unit. Using the same example as before, we break the fraction into a constant stack, an x-stack, and a y-stack. Then we simplify each stack. Then we combine the stacks to make a new fraction.

$$\frac{756x}{72xy} = \left(\frac{756}{72}\right)\left(\frac{x}{x}\right)\left(\frac{1}{y}\right) = \left(\frac{21}{2}\right)\left(\frac{1}{1}\right)\left(\frac{1}{y}\right) = \frac{21}{2y}$$

Problems on distributing and factoring (unit 3.4)

1. The expression $3xy\left(2x^2 + xy\right)$ is equivalent to:

   (A) $7x^2y$

   (B) $7x^3y$

   (C) $9x^2y$

   (D) $6x^3y + xy$

   (E) $6x^3y + 3x^2y^2$

2. The expression $3xy^2 - 18x^3y^3$ is equivalent to:

   (A) $3xy^2\left(1 - 6x^2y\right)$

   (B) $3xy(1 - 6xy)$

   (C) $3x^2y^2(1 - 6xy)$

   (D) $xy(3y - 18xy)$

   (E)   None of the above

3. The expression $648x^3y^6 - 162xyz$ is equivalent to:

   (A) $81xy(8x^2y^5 - z)$

   (B) $81xyz(8x^2y^5 - 1)$

   (C) $162xy(4x^2y^5 - z)$

   (D) $162xyz(4x^2y^5 - 1)$

   (E)   None of the above

4. The simplest form of the expression
$360x^8y^5z^2 - 480x^3y^2z^3 + 540x^5y^7z^8$ is:

   (A) $15x^3y^2z\left(24x^5y^3z - 32z^2 + 36x^2y^5z^5\right)$

   (B) $60x^3y^2z\left(6x^5y^3z - 8z^2 + 9x^2y^5z^5\right)$

   (C) $15x^3y^2z^2\left(24x^5y^3 - 32z + 36x^2y^5z^6\right)$

   (D) $60x^3y^2z^2\left(6x^5y^3 - 8z + 9x^2y^5z^6\right)$

   (E)   None of the above

5. The expression $5xy$ is equivalent to:

   (A) $\dfrac{50x^2y^2}{5xy}$

   (B) $\dfrac{30x^2y}{6x}$

   (C) $\dfrac{12x^3y}{60x^2}$

   (D) $\dfrac{15x^3y^8}{3x^4y^7}$

   (E)   None of the above

6. The expression $\dfrac{120x^3y^2 - 136x^2z}{8x^2}$ can be simplified to:

   (A) $15xy^2 - 17z$

   (B) $15y^2 - 17z$

   (C) $-2xyz$

   (D) $-2(xy^2 - 17)$

   (E) $\dfrac{15y^2}{x} - \dfrac{17z}{x^2}$

7. The expression $\dfrac{96xy^2z^3 - 36x^2yz^2}{12x^3yz^6}$ can be simplified to:

   (A) $\dfrac{8yz - 3x}{xz^2}$

   (B) $8yz - 3x$

   (C) $\dfrac{8y}{x^2} - \dfrac{3}{xz^4}$

   (D) $\dfrac{8yz - 3x}{x^2z^4}$

   (E) $\dfrac{8}{x^2} - \dfrac{3}{xz^4}$

Solutions to problems on distributing and factoring (unit 3.4)

1. (E)

$$3xy\left(2x^2 + xy\right) = 3xy \cdot 2x^2 + 3xy \cdot xy = 6x^3 y + 3x^2 y^2$$

2. (A)

$$3xy^2 - 18x^3 y^3 = 3xy^2\left(1 - 6x^2 y\right)$$

3. (C)

Find GCF of 648 and 162:

$$648 = 2^3 \cdot 3^4 \quad \text{and} \quad 162 = 2 \cdot 3^4 \quad \text{so GCF} = 2 \cdot 3^4 = 162$$
$$648x^3 y^6 - 162xyz = 162xy\left(4x^2 y^5 - z\right)$$

4. (D)

Use factor trees to find the GCF of 360 480 and 540

$$360 = 2^3 \cdot 3^2 \cdot 5 \quad 480 = 2^5 \cdot 3 \cdot 5 \quad 540 = 2^2 \cdot 3^3 \cdot 5$$
$$\text{so } \text{GCF} = 2^2 \cdot 3 \cdot 5 = 60$$
$$360x^8 y^5 z^2 - 480x^3 y^2 z^3 + 540x^5 y^7 z^8 =$$
$$60x^3 y^2 z^2\left(6x^5 y^3 - 8z + 9x^2 y^5 z^6\right)$$

5. (B)

$$\frac{30x^2 y}{6x} = \frac{6x\left(5xy\right)}{6x} = 5xy$$

6. (A)

Use your calculator or factor trees to find GCF of 120 and 136

$$120 = 2^3 \cdot 3 \cdot 5 \quad \text{and} \quad 136 = 2^3 \cdot 17 \quad \text{so GCF} = 2^3 = 8$$

$$\frac{120x^3 y^2 - 136x^2 z}{8x^2} = \frac{8x^2\left(15xy^2 - 17z\right)}{8x^2} = 15xy^2 - 17z$$

7. (D)

Use factor trees or calculator for the GCF of 96 and 36

$$96 = 2^5 \cdot 3 \quad \text{and} \quad 36 = 2^2 \cdot 3^2 \quad \text{so GCF} = 2^2 \cdot 3 = 12$$

$$\frac{96xy^2 z^3 - 36x^2 yz^2}{12x^3 yz^6} = \frac{12xyz^2\left(8yz - 3x\right)}{12x^3 yz^6} = \frac{8yz - 3x}{x^2 z^4}$$

Unit 3.5 FOILing and factoring

The acronym FOIL reminds us how to multiply terms that look like

$$(a+b)(c+d) = a(c+d)+b(c+d) = ac + ad + bc + bd$$

The acronym comes from

F	= first (the first terms are a and c)
O	= outer (the outer terms are a and d)
I	= inner (the inner terms are b and c)
L	= last (the last terms are b and d)

That's all there is to it! For example

$$(2x+3)(x-y) = 2x^2 - 2xy + 3x - 3y$$

First	$2x \cdot x = 2x^2$
Outer	$2x(-y) = -2xy$
Inner	$3 \cdot x = 3x$
Last	$3(-y) = -3y$

Factoring

Factoring is just FOILing in reverse. In fact, one of my students used to call factoring "reverse FOILing." The general problem is that we want to factor the expression

$$ax^2 + bx + c .$$

There are two cases to consider. The easy case is where a=1, so let's start there.

Factoring when $a = 1$

When a=1, the equation looks like

$$x^2 + bx + c$$

There are three steps to factoring this:

1. Begin with the setup of $(x \quad)(x \quad)$.
2. Find all of the pairs of factors of c. Then find a pair for which the sum or difference equals b. Enter that pair into the setup.
3. Adjust the signs.

Consider the following example:

$$x^2 - 9x - 10$$

1. Begin with the setup $(x \quad)(x \quad)$.
2. The factors of 10 are 2x5 and 1x10. 2 and 5 will not work because 2+5=7 and 5-2=3. We are looking for 9, not 7 or 3. Next try 1 and 10. 1+10=11 but 10-1=9. So we know to use 1 and 10. We fill the setup as follows: $(x \quad 1)(x \quad 10)$.
3. We know we need a plus sign and a negative sign because 10 is negative. The negative must go with the 10 and not the 1 because the 9 is negative, so
$$x^2 - 9x - 10 = (x+1)(x-10).$$

Always confirm your answer by FOILING:

$$(x+1)(x-10) = x^2 - 10x + x - 10 = x^2 - 9x - 10.$$

Factoring when $a \neq 1$

Now consider the more difficult case of

$$ax^2 + bx + c.$$

The steps are:

1. Create a setup for each pair of factors of a.
2. For each setup follow steps 1-3 above.

Consider factoring the following:

$$4x^2 + 5x - 6$$

Two setups will be needed, one for 2 and 2; and another for 4 and 1. The setup for 2 and 2 looks like $(2x \quad)(2x \quad)$. For this setup we can try factors of 6, namely 6 and 1; and 2 and 3. The combination of 6 and 1 does not work because 12 and 2 neither add or subtract to give 5. The combination of 2 and 3 does not work because 4 and 6 neither add or subtract to give 5. How annoying! So next we try the setup for 4 and 1, which looks like $(4x \quad)(x \quad)$. Now we try factors of 6, starting with 6 and 1. That combination does not work because neither 24 and 1, nor 4 and 6 will give us the 5 that we need. The combination of 2 and 3 can work because 8-3 equals 5. So the answer is

$$4x^2 + 5x - 6 = (4x - 3)(x + 2).$$

Factoring is not difficult but it can be a nuisance and it does require practice.

Special cases (perfect squares)

There are three patterns of factoring and FOILing that you must memorize and be able to recognize on the test. These pertain to perfect squares:

$$(x+y)^2 = x^2 + 2xy + y^2$$
$$(x-y)^2 = x^2 - 2xy + y^2$$
$$(x+y)(x-y) = x^2 - y^2$$

It is not good enough to be able to FOIL or factor these from scratch when you encounter them on the test. This sample problem illustrates why.

Suppose $(x+y)^2 = 100$, $x^2 = 20$, and $y^2 = 30$. Find the value of xy. If you do not recognize the pattern you will not know what to do. However recognizing the pattern enables the problem to be solved easily.

$$(x+y)^2 = x^2 + 2xy + y^2$$
$$100 = 20 + 2xy + 30$$
$$50 = 2xy$$
$$25 = xy$$

Problems on FOILing and factoring (unit 3.5)

1. The expression $(2x-3)(5x+7)$ is equivalent to:

   (A) $7x+4$

   (B) $-3x-10$

   (C) $10x^2+14$

   (D) $10x^2-x-21$

   (E) $10x^2+29-21$

2. The expression $(x+8)^2$ is equivalent to:

   (A) $2x+16$

   (B) $x^2+8x+64$

   (C) x^2+64

   (D) x^2-64

   (E) $x^2+16x+64$

3. The expression x^2-6x+8 is equivalent to:

   (A) $(x+8)(x-1)$

   (B) $(x-8)(x+1)$

   (C) $(x+4)(x+2)$

   (D) $(x+4)(x-2)$

   (E)   None of the above

4. The expression $x^2-5xy-176y^2$ is equivalent to:

   (A) $(x-16y)(x+11y)$

   (B) $(x-8y)(x-22y)$

   (C) $(x-8y)(x+22y)$

   (D) $(x+8y)(x-22y)$

   (E) $(x+8y)(x+22y)$

5. The expression $3x^2-19x+20$ is equivalent to:

   (A) $(3x-5)(x-4)$

   (B) $(3x-5)(x+40)$

   (C) $(3x-4)(x-5)$

   (D) $(3x-4)(x+5)$

   (E)   None of the above

6. The expression $20x^2-36xy+16y^2$ is equivalent to:

   (A) $(5x+8y)(4x+8y)$

   (B) $(5x-8y)(4x+2y)$

   (C) $(2x+y)(10x-16y)$

   (D) $4(x-y)(5x-4y)$

   (E) $(20x+8y)(x-2y)$

7. The expression $\dfrac{2x^2-6x-8}{x^2-1}$ is equivalent to:

   (A) -8

   (B) $\dfrac{2x-8}{x-1}$

   (C) $\dfrac{2x+8}{x+1}$

   (D) $\dfrac{2x-8}{x+1}$

   (E)   None of the above

8. If $x^2-y^2=50$ and $x+y=20$, what is the value of $x-y$?

   (A)   0

   (B)   1

   (C)   2

   (D)   2.5

   (E)   Cannot be determined

Solutions to problems on FOILing and factoring (unit 3.5)

1 (D)

$$(2x-3)(5x+7) = 10x^2 + 14x - 15x - 21$$
$$= 10x^2 - x - 21$$

2. (E)

$$(x+8)^2 = (x+8)(x+8) = x^2 + 8x + 8x + 64$$
$$= x^2 + 16x + 64$$

3. (E)

$$x^2 - 6x + 8 = (x-4)(x-2)$$
$$= x^2 - 4x - 2x + 8 = x^2 - 6x + 8$$

4. (A)

$$x^2 - 5xy - 176y^2 = (x - 16y)(x + 11y)$$
$$= x^2 - 16xy + 11xy - 176 = x^2 - 5xy - 176$$

5. (C)

$$3x^2 - 19x + 20 = (3x - 4)(x - 5)$$
$$= 3x^2 - 15x - 4x + 20 = 3x^2 - 19x + 20$$

6. (D)

$$20x^2 - 36xy + 16y^2 = 4(5x^2 - 9xy + 4y^2)$$
$$= 4(x - y)(5x - 4y)$$

7. (B)

$$\frac{2x^2 - 6x - 8}{x^2 - 1} = \frac{2(x^2 - 3x - 4)}{(x+1)(x-1)} = \frac{2(x+1)(x-4)}{(x+1)(x-1)}$$
$$= \frac{2(x-4)}{x-1} = \frac{2x-8}{x-1}$$

8. (D)

$$x^2 - y^2 = (x+y)(x-y)$$
$$50 = 20(x - y)$$
$$2.5 = x - y$$

Unit 3.6 Ratios

Ratios are very useful things to know. But before they can be applied, it is necessary to be able to solve them by cross-multiplying. For example:

$$\frac{3}{5} = \frac{12}{x}$$
original problem

$$3x = 5 \cdot 12$$
cross-multiply

$$3x = 60$$
simplify

$$x = 20$$
solve for x by dividing both sides by 3.

Ratios can be written as fractions (as done above) or they can be written using the : symbol. For example, the problem above could be written 3:5 as 12:x. Or a problem might use words, as in "3 is to 5 as 12 is to x."

Unit Conversions

Ratios are especially useful for unit conversions. For example, suppose we want to know how many feet there are in 84 inches. We can find the answer using ratios:

$$\frac{12}{1} = \frac{84}{x}$$
$$12x = 84$$
$$x = 7$$

The first ratio comes from the fact that there are 12 inches in a foot. The second ratio is setup so that the inches amount is in the numerator and the unknown variable for the feet amount is in the denominator. When you setup ratios to do unit conversion, make sure that your units match up -- the same units should be in both numerators and the same units should be in both denominators.

Mixture Problems

There is a type of problem called mixture problems and they frequently show up on standardized tests. At first they look difficult but once you see how to solve them, they are easy. Consider the following:

> You are making pizza dough according to the recipe of 9 parts flour to 2 parts water. You need to make 165 pounds of dough. How many pounds of flour and how many pounds of water should you use?

This seems difficult at first, but the solution is very easy. We just set up the equation

$$9x + 2x = 165$$

What this equation says is that the same multiplier (x) will be applied to the 9 parts of flour and the 2 parts of water, and that they must sum to 165. Solving for x gives us 15. So we must use $9 \cdot 15 = 135$ pounds of flour and $2 \cdot 15 = 30$ pounds of water. Notice that 135 and 30 sum to the 165 pounds of dough required.

Direct and Inverse Variation

This is another type of word problem that seems difficult until you see how to do it. We begin with the definitions:

The variables x and y vary directly if their relationship can be modeled as

$$y = kx$$

where k is a non-zero constant. If we suppose k to be positive, you can see that when the value of x increases, y increases. This is where the name "direct" variation comes from. The variables x and y move in the same direction.

The variables x and y vary inversely if their relationship can be modeled as

$$y = \frac{k}{x}$$

where k is a non-zero constant. If we suppose k to be positive, you can see that when the value of x increases, y decreases. This is where the name "indirect" variation comes from. The variables x and y move in opposite directions.

These problems usually take the same form. They begin by saying whether the relationship is direct or inverse and supply initial values of x and y. This enables you to evaluate k. Then the problem asks for a new value of y when the value of x changes. For example:

   X and Y are inversely proportional, and when X=10 Y=20. What is the value of Y when X=5?

The problem is solved as follows:

$$y = \frac{k}{x}$$ given

$$20 = \frac{k}{10}, \quad 200 = k$$ substitute and solve for k

$$y = \frac{200}{x}$$ new partial equation

$$y = \frac{200}{5} = 40$$ substitute for X, solve for Y.

Note that Y increased when X decreased because X and Y are inversely related.

Problems on ratios (unit 3.6)

1. Solve for x $\dfrac{5}{6x} = \dfrac{10}{7}$, $x \neq 0$

(A) $\dfrac{7}{12}$

(B) $\dfrac{12}{7}$

(C) $\dfrac{25}{21}$

(D) $\dfrac{21}{25}$

(E) Cannot be determined

2. Solve for x: $\dfrac{4}{x} = \dfrac{x}{16}$, $x \neq 0$

I. 1/4

II. 8

III. -8

(A) I only

(B) II only

(C) III only

(D) I and II only

(E) II and III only

3. You sell mixed nuts over the Internet. Your secret recipe is 9 parts peanuts to 3 parts cashews to 1 part pecans. How many pounds of cashews will be needed to fill an order for 195 pounds of mixed nuts?

(A) 15

(B) 38

(C) 45

(D) 49

(E) 135

4. The interior angles of a quadrilateral are in the ratio of 4:3:2:1. What is the size of the largest angle?

(A) 18

(B) 36

(C) 72

(D) 108

(E) 144

5. If there are 16 ounces in a pound, how many pounds are in 80 ounces?

(A) 1/5

(B) 5

(C) 64

(D) 96

(E) 1280

6. A blueprint is drawn to scale, where two inches represents 25 feet. If a room is 70 feet long, how many inches long will it be on the blueprint?

(A) 0.2

(B) 0.7

(C) 1.4

(D) 5.6

(E) 875

7. Y varies directly with X. When X=5, Y=40. What is the value of Y when X increases to 7?

(A) 7/8

(B) 8/7

(C) 15

(D) 25

(E) 56

8. Y varies inversely with X. When X=5, Y=40. What is the value of Y when X increases to 7?

(A) 7/8

(B) 8/7

(C) 25

(D) 200/7

(E) 1600

Solutions to problems on ratios (unit 3.6)

1. (A)

$$\frac{5}{6x} = \frac{10}{7}, \quad 60x = 35, \quad x = \frac{35}{60} = \frac{7}{12}$$

2. (E)

$$\frac{4}{x} = \frac{x}{16}, \quad x^2 = 64, \quad x = \pm 8$$

3. (C)

$$9x + 3x + x = 195$$
$$13x = 195$$
$$x = 15, \quad 3x = 45 \text{ pounds of cashews}$$

4. (E)

The interior angles sum to 360.

$$4x + 3x + 2x + x = 360$$
$$10x = 360$$
$$x = 36, \quad 4x = 144 = \text{the largest angle}$$

5. (B)

$$\frac{16}{1} = \frac{80}{x}, \quad 16x = 80, \ x = 5 \text{ pounds}$$

6 (D)

$$\frac{2}{25} = \frac{x}{70}, \quad 25x = 140, \quad x = 5.6 \text{ inches}$$

7. (E)

$$y = kx$$
$$40 = 5k, \quad k = 8$$
$$y = 8x, \quad y = 8 \cdot 7 = 56$$

8 (D)

$$y = \frac{k}{x}$$
$$40 = \frac{k}{5}, \quad k = 200$$
$$y = \frac{200}{x}, \quad y = \frac{200}{7}$$

4. GEOMETRY

Geometry uses different circuitry in your brain from that used by algebra. If you have not done geometry in a while you will probably need to get those wheels turning again.

Unit 4.1 Angles and lines

There are three types of angles to know:

$$0° < \text{acute angle} < 90°$$

$$\text{right angle} = 90°$$

$$90 < \text{obtuse angle} < 180°$$

Supplementary angles comprise a straight line (they sum to 180 degrees); whereas complementary angles comprise a right angle (they sum to 90 degrees).

Supplementary   Complementary

Remember that vertical angles are congruent (they have the same measure).

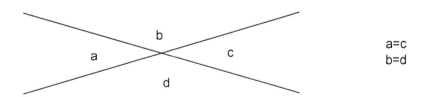

a=c
b=d

The bulk of what you need to remember about angles and lines has to do with parallel lines with a transversal line. The picture looks like this:

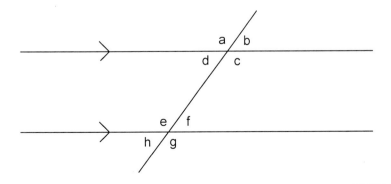

There is a bit of terminology to be mastered with this. The interior angles are those angles that lie inside the parallel lines, namely c, d, e and f. The exterior angles are those that lie outside the parallel lines, namely a, b, h and g.

There are several sets of congruent pairs to remember, as you can see in the table below.

Vertical angles	a=c
	b=d
	e=g
	f=h
Alternate interior angles	c=e
	d=f
Corresponding angles	a=e
	d=h
	b=f
	c=g
Alternate exterior angles	a=g
	b=h

Problems on angles and lines (unit 4.1)

Questions 1 and 2 refer to the diagram below.

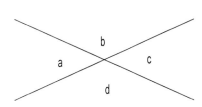

1. Which of the following statements must be true?

    I.   a=c

    II. a=d

    III. a+b=c+d

(A)   I only

(B)   II only

(C)   III only

(D)   I and II only

(E)   None of the above

2. If a:b:c:d as 4:3:2:1 then what is the measure of angle B?

(A)   6 degrees

(B)   24 degrees

(C)   36 degrees

(D)   54 degrees

(E)   108 degrees

3. If A and B are complementary angles and the measure of A is 20 degrees, what is the measure of B?

(A)   70 degrees

(B)   90 degrees

(C)   160 degrees

(D)   180 degrees

(E)   340 degrees

Questions 4 and 5 refer to the diagram below.

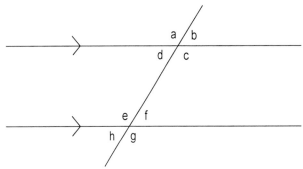

4. Which of the following statements must be true?

    I.   a=c

    II. a=h

    III. a+c=e+g

(A)   I only

(B)   II only

(C)   III only

(D)   I and II only

(E)   I and III only

5. Which of the following statements is not necessarily true?

    I.   a=g

    II. b=e

    III. d=h

(A)   I only

(B)   II only

(C)   III only

(D)   I and II only

(E)   II and III only

6 If a and b are supplementary angles, and a and c are complementary angles, then:

(A)   a+b+c = 270 degrees

(B)   b - a = 90 degrees

(C)   b −c = 90 degrees

(D)   a+b-c = 90 degrees

(E)   b + c = 180 degrees

Solutions to problems on angles and lines (unit 4.1)

1 (E)

None of the above.

I is true (a and c are vertical angles), II is not necessarily true, and III is true (a+b=180=c+d). I and III is not one of the choices.

2. (E).

This is a mixture problem:

$$4x + 3x + 2x + x = 360$$
$$10x = 360, \quad x = 36$$
$$3x = 108$$

3. (A).

$$A + B = 90, \quad B = 90 - 20 = 70$$

4. (E)

I is true because a and c are vertical. II is false because a and h are not alternate exterior. III is true because a=c and e=g (vertical) <u>and</u> c=e (alternate interior). Therefore a=c=e=g and a+c=e+g.

5. (B)

I is always true because a and g are alternate exterior angles. II may not be true. III is always true because d and h are corresponding angles.

6. (C)

This is a pair of equations:

$$a + b = 180$$
$$a + c = 90$$
$$b - c = 180 - 90 = 90$$

Unit 4.2 Triangles

Triangles are a big part of standardized tests, especially right triangles. But first we begin with information about triangles in general. There are three types of triangles:

	Sides	Angles
Equilateral	All congruent	All 60 degrees
Isosceles	Two congruent	Two congruent
Scalene	None congruent	None congruent

The formulas for the area of a triangle are:

Any triangle	$A = \dfrac{1}{2}bh$
Right triangle	$A = \dfrac{1}{2}leg_1 \cdot leg_2$
Equilateral triangle	$A = \dfrac{s^2\sqrt{3}}{4}$

Triangle Inequality

For reasons that are unclear to me, the standardized tests are fond of the triangle inequality. It simply states that any leg of any triangle is strictly less than the sum of the other two legs. Graphically, it looks like this:

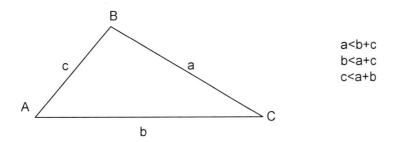

Note that it is only necessary to test whether the longest leg is strictly less than the sum of the other two legs. In diagram above, it is only necessary to test whether b is less than $a+c$. Using a variety of means, a question may ask whether a particular triplet can form a triangle, for example 1,2,3. That cannot be a triangle because 3=1+2. However 1,2,2 can be a triangle because 2<1+2.

Along with the triangle inequality, remember that the longest side (b in the diagram) is opposite the largest angle (B in the diagram); whereas the shortest side (c in the diagram) is opposite the smallest angle (C in the diagram). A corollary to that is the hypotenuse of a right triangle is the longest side (and it is opposite the largest angle, the right angle).

Similar Triangles

Most students remember congruent triangles, where corresponding angles are congruent and corresponding sides are congruent (remember SSS, SAS, AAS, and ASA?). However, there is a tendency to forget similar triangles, where corresponding angles are congruent and corresponding sides are proportional. Similar triangles are illustrated below:

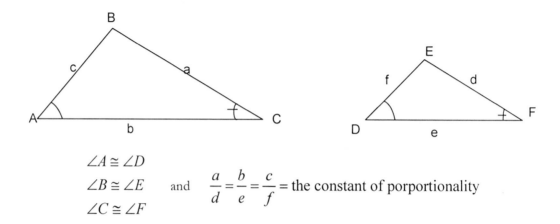

$$\angle A \cong \angle D$$
$$\angle B \cong \angle E \qquad \text{and} \qquad \frac{a}{d} = \frac{b}{e} = \frac{c}{f} = \text{the constant of porportionality}$$
$$\angle C \cong \angle F$$

A special case of similar triangles is where a line has been drawn through a triangle so that the line is parallel to one of the sides of the triangle, as shown below. The result is two similar triangles, where the larger original triangle is similar to the smaller triangle.

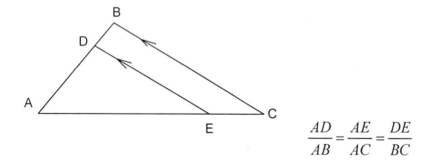

$$\frac{AD}{AB} = \frac{AE}{AC} = \frac{DE}{BC}$$

Note that angle ADE and angle ABC are congruent because they are corresponding angles. The same is true for angles AED and ACB. Having proved angles to be congruent we can conclude that the triangles are similar (remember AA?).

Right Triangles

The most famous theorem in all of math, the Pythagorean Theorem, applies to right triangles,. The theorem states

$$a^2 + b^2 = c^2$$

where c is the length of the hypotenuse, and a and b are the lengths of the legs.

It is absolutely necessary to memorize the relationships of the sides of the two special triangles shown below, even though these are supplied on the SAT®. You must recognize these relationships when they show up on the test, without flipping pages back and forth when a formula is needed.

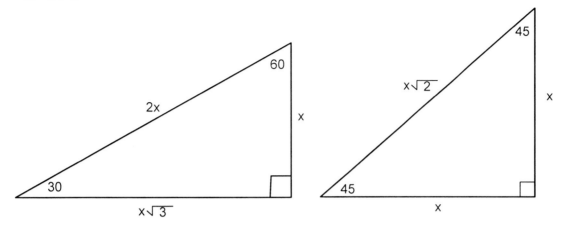

Be on the lookout for these special right triangles because they are very popular on standardized tests. The table below gives some tips on how to spot them.

When you see:	Look out for:
$\sqrt{3}$	30-60-90
A side that is $\dfrac{1}{2}$ the hypotenuse	30-60-90
A hypotenuse that is twice the length of a side	30-60-90
An equilateral triangle	30-60-90
$\sqrt{2}$	45-45-90
The diagonal of a square	45-45-90

A useful exercise is to use the Pythagorean Theorem to prove the relationships of the sides of each of the special right triangles. The 45-45-90 triangle is shown below as an example, and the 30-60-90 triangle is left as an exercise.

$$x^2 + x^2 = \left(x\sqrt{2}\right)^2$$
$$2x^2 = \sqrt{2} \cdot \sqrt{2} \cdot x \cdot x = 2x^2$$

Practice problems on triangles (unit 4.2)

1. What is the perimeter of an equilateral triangle whose area is $4\sqrt{3}$?

(A)   2

(B)   4

(C)   8

(D)   12

(E)   16

2. What is the side of an equilateral triangle whose altitude is 3?

(A)   1.5

(B) $\sqrt{3}$

(C) $2\sqrt{3}$

(D)   3

(E)   6

3. What is the area of a square whose diagonal is 2?

(A)   -2

(B)   1

(C)   2

(D) $\sqrt{2}$

(E)   4

4. One leg of a right triangle is 5 and its hypotenuse is 10. What is the size of its smallest angle?

(A)   15 degrees

(B)   30 degrees

(C)   45 degrees

(D)   60 degrees

(E)   75 degrees

5. One leg of a right triangle is 7 and the hypotenuse is 10. What is the length of the other leg?

(A)   6

(B)   9

(C) $\sqrt{51}$

(D)   36

(E) $\sqrt{149}$

6. One side of a triangle is 7 and another side is 10. Which of the values below could be the third side of the triangle?

I.   2   IV 16

II. 3   V. 17

III. 4   VI. 18

(A)   I and IV only

(B)   II and V only

(C)   III and IV only

(D)   I, II and III only

(E)   IV, V, and VI only

7. In the diagram below, triangle ABC is similar to triangle DEF. What is the value of y-x?

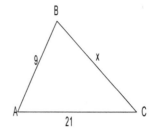 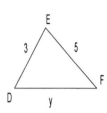

(A)   -53

(B)   -8

(C)   5/3

(D)   7

(E)   -63

8. A six foot tall man stands in the shadow of a 18 foot tall telephone pole. If the man is standing three feet from the base of the telephone pole, what is the length of the telephone pole's shadow?

(A)   3

(B)   3.5

(C)   4

(D)   4.5

(E)   Cannot be determined

Solutions to practice problems on triangles (unit 4.2)

1. (D)

$$A = s^2 \frac{\sqrt{3}}{4}$$

$$4\sqrt{3} = \frac{s^2 \sqrt{3}}{4}$$

$$s^2 = 16, \quad s = 4, \quad P = 12$$

2. (C)

The altitude of an equilateral triangle forms two 30-60-90 triangles. The altitude is opposite the 60 degree angle, and the side opposite the 30 degree angle is s/2. So if the altitude is 3 then

$$3 = \frac{s}{2}\sqrt{3}, \quad \frac{6}{\sqrt{3}} = s, \quad 2\sqrt{3} = s.$$

3. (C)

The diagonal of a square forms two 45-45-90 triangles, with the diagonal being the hypotenuse of each right triangle. So if the diagonal is 2 then

$$s^2 + s^2 = 2^2, \ 2s^2 = 4, \ s=\sqrt{2}$$

$$A = s^2 = 2$$

4. (B)

This must be a 30-60-90 triangle because the hypotenuse is twice as long as of one of the sides. Therefore the smallest angle is 30 degrees.

5. (C)

$$x^2 + 7^2 = 10^2$$

$$x^2 = 100 - 49 = 51$$

$$x = \sqrt{51}$$

6. (C)

I. False $10 \ not < \ 2+7$

II False $10 \ not < \ 3+7$

III. True $10 < \ 4+7$

IV True $16 < \ 7+10$

V. False $17 \ not < \ 7+10$

VI. False $18 \ not < \ 7+10$

7. (B)

$$\frac{3}{9} = \frac{y}{21}, \quad 9y = 63, \quad y = 7$$

$$\frac{3}{9} = \frac{5}{x}, \quad 3x = 45, \quad x=15$$

$$y - x = 7 - 15 = -8$$

8. (D)

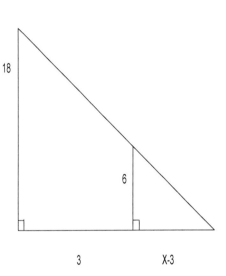

There are two similar right triangles, where the base of the large triangle is x and the base of the small triangle is x-3. Use the ratios of little over big to find

$$\frac{x-3}{x} = \frac{6}{18}$$

$$18x - 54 = 6x$$

$$12x = 54$$

$$x = 4.5$$

Unit 4.3 Circles

There are just a few things to remember about circles, and most of them are things you probably know already. Everyone remembers that the central angle of an entire circle measures 360 degrees and that the measure of the entire arc of a circle is 360 degrees. But you must also remember the relationship between arcs, central angles, and inscribed angles, as shown below.

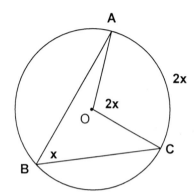
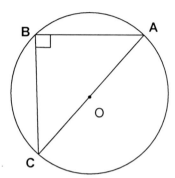

Take a look at circle O on the left. It has a central angle AOC (angle AOC is a central angle because its vertex is the center of the circle). Note that the measure of central angle AOC is the same as the measure of its inscribed arc, arc AC. On the other hand, angle ABC is an inscribed angle, not a central angle (its vertex is not at the center of the circle). Its measure is one-half the measure of its inscribed arc.

Circle O on the right illustrates a special case of an inscribed angle. Angle ABC is inscribed in a semi-circle, where chord AC is a diagonal of the circle. Therefore arc AC is a semi-circle and must be 180 degrees (half of a 360 degree circle). Because angle ABC is inscribed in a semi-circle, its measure must be 90 degrees (half of the 180 degree arc in which it is inscribed). For some reason, inscribed right angles are popular questions.

Another favorite topic has to do with sectors of circles. Think of these as slices from a pizza pie. You need to memorize formulas for the area of the pizza slice (area of the sector) and the length of the crust of the pizza slice (length of the arc). These are shown below.

Notice that the central angle, which is x degrees, is the key to understanding how these formulas work. As x increases, the size of the slice increases. If the slice were half of the pizza, x would be 180 degrees. The formulas simply take the fraction of the whole that the central angle represents and then multiply that fraction by the area of the whole pie (πr^2) or by the circumference of the whole pie ($2\pi r$).

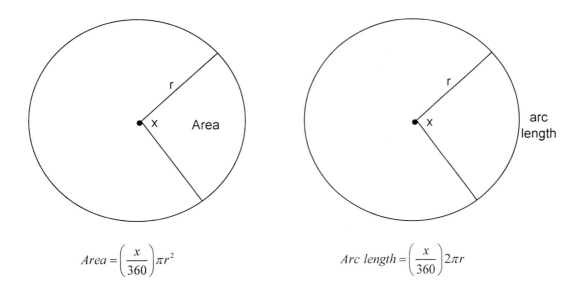

$$Area = \left(\frac{x}{360}\right)\pi r^2 \qquad\qquad Arc\ length = \left(\frac{x}{360}\right)2\pi r$$

Lastly, it is necessary to remember that the tangent to a circle forms a right angle with the radius of a circle at the point where the tangent and radius meet. This is illustrated below.

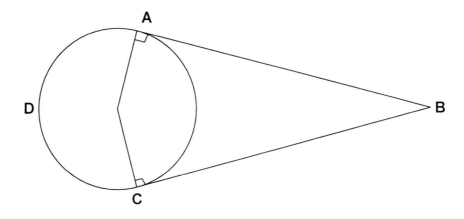

Practice problems on circles (unit 4.3)

1. If circle A has a radius that is twice the length of the radius of circle B, then the following must be true:

      I. The area of circle A is two times the area of circle B.

      II. The circumference of circle A is two times the circumference of circle B.

      III. The area of circle A is four times the area of circle B.

    (A)   I only

    (B)   II only

    (C)   III only

    (D)   I and II only

    (E)   II and III only

2. In the figure below, which of the following must be true?

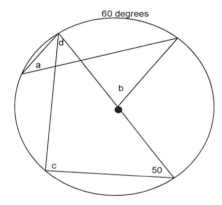

60 degrees

      I. a=60

    II. c-d=50

    III. b=120

    (A)   I only

    (B)   II only

    (C)   III only

    (D)   I and II only

    (E)   None of the above

3. If a circle is divided into eight congruent sectors and each sector has an area of 8π, what is the radius of the circle?

    (A)   4

    (B)   8

    (C)   12

    (D)   18

    (E)   32

4. If a sector has an arc length of π and a radius of 10, what is the size of its central angle?

    (A)   4

    (B)   8

    (C)   12

    (D)   18

    (E)   20

5. Billy ordered a pizza with a crust that is 10π inches long. What is the area of the pizza?

    (A) 5π

    (B) 10π

    (C) 15π

    (D) 20π

    (E) 25π

6. A tangent is drawn to a circle from an external point that is 10 inches from the center of the circle. If the external point is 20 inches from the point of tangency, what is the radius of the circle?

    (A)   10

    (B) $10\sqrt{3}$

    (C) $10\sqrt{5}$

    (D)   20

    (E)   Cannot be determined

Solutions to practice problems on circles (unit 4.3)

1 (E)

I is false and III is true:

$$\frac{A_A}{A_B} = \frac{\pi r_A^2}{\pi r_B^2} = \frac{(2r_B)^2}{r_B^2} = 4$$

II is true:

$$\frac{C_A}{C_B} = \frac{2\pi r_A}{2\pi r_B} = \frac{r_A}{r_B} = \frac{2r_B}{r_B} = 2$$

2. (B)

I is false; a=30, inscribed angle

II is true because c=90 and therefore d=40

III is false because b=60, central angle.

3. (B)

$$8\pi = \left(\frac{45}{360}\right)\pi r^2, \ 8 = \frac{1}{8}r^2, \ 64 = r^2, \ 8 = r$$

4. (D)

$$\pi = \left(\frac{x}{360}\right)2\pi \cdot 10, \ 1 = 20\left(\frac{x}{360}\right)$$

$$\frac{1}{20} = \frac{x}{360}, \quad x = 18$$

5. (E)

$$2\pi r = 10\pi, \ r = 5$$
$$A = \pi r^2 = 25\pi$$

6. (B)

$$r^2 + 10^2 = 20^2$$
$$r^2 = 400 - 100 = 300$$
$$r = \sqrt{300} = 10\sqrt{3}$$

Unit 4.4 Polygons

A polygon is a closed shape with n straight sides. If n=3 the polygon is a triangle. If n=4 the polygon is a quadrilateral, etc. A **regular polygon** has all sides congruent and all interior angles congruent. There are two formulas to be memorized for polygons

$$\text{sum of interior angles} = (n-2)180$$

$$\text{sum of exterior angles} = 360$$

The derivation of the first formula above is interesting and kind of fun to do. Draw a polygon and then choose a vertex at random. From that vertex draw lines to all of the other vertices in the polygon. You should now have divided the polygon into $(n-2)$ triangles. As each triangle contains angles that sum to 180 degrees, the sum of the interior angles of the polygon is $(n-2)180$.

Quadrilaterals

The most popular polygon is the four-sided polygon or quadrilateral. There are quite a few properties to remember about quadrilaterals, and the easiest way to remember them is to use a "quadrilateral tree," as shown below.

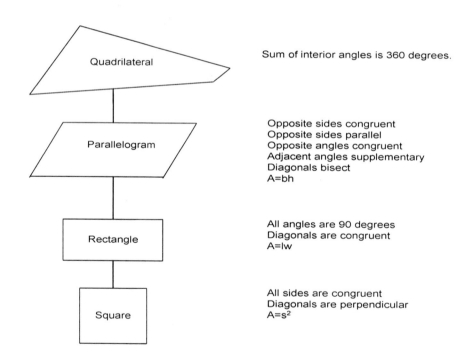

Quadrilateral — Sum of interior angles is 360 degrees.

Parallelogram — Opposite sides congruent / Opposite sides parallel / Opposite angles congruent / Adjacent angles supplementary / Diagonals bisect / A=bh

Rectangle — All angles are 90 degrees / Diagonals are congruent / A=lw

Square — All sides are congruent / Diagonals are perpendicular / A=s²

The key to understanding the tree is that properties are inherited by the "children" from the "parents." The parallelogram is a special type of quadrilateral, the rectangle is a special type of parallelogram, and the square is a special type of rectangle. For example, the diagonals of a rectangle bisect each other because that property is inherited from the parallelogram.

Practice problems on polygons (unit 4.4)

1. How large (in degrees) is the interior angle of a regular pentadecagon (15 sides)?

    (A)   40

    (B)   66

    (C)   110

    (D)   125

    (E)   156

2. How large (in degrees) is the exterior angle of a regular dodecagon (12 sides)?

    (A)   15

    (B)   30

    (C)   45

    (D)   60

    (E)   90

3. If sum of the interior angles of a regular polygon is 1800 degrees, how many sides does it have?

    (A)   6

    (B)   8

    (C)   10

    (D)   12

    (E)   14

4. In the parallelogram below, what is the measure of z (in degrees)?

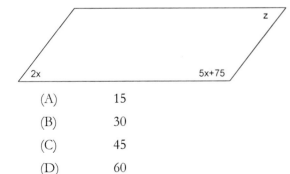

    (A)   15

    (B)   30

    (C)   45

    (D)   60

    (E)   90

5. If a rectangle has an area of 60 units and a width of 5 units, what is the length of its diagonal?

    (A) $\sqrt{61}$

    (B)   12

    (C)   13

    (D) $\sqrt{119}$

    (E)   Cannot be determined

6. In the parallelogram below, what is the measure (in degrees) of z?.

    (A)   14.7

    (B)   31

    (C)   63.2

    (D)   116.8

    (E)   149

7. If the base of a parallelogram is doubled and its height is tripled, what is the ratio of the area of the new parallelogram to the old parallelogram?

    (A)   1

    (B)   2

    (C)   3

    (D)   6

    (E)   36

8. If the area of a square is 10 square units, what is the sum of its diagonals?

    (A) $2\sqrt{5}$

    (B) $4\sqrt{5}$

    (C) $2\sqrt{10}$

    (D) $4\sqrt{10}$

    (E)   20

Solutions to practice problems on polygons (unit 4.4)

1. (E)

Interior angles sum to $(15-2)180 = 2340$.

Each interior angle must be $2340/15 = 156$ degrees.

2. (B)

Each exterior angle must be $360/12 = 30$ degrees.

3. (D)

$$1800 = (n-2)180$$
$$10 = n-2, \quad n = 12 \text{ sides}$$

4. (B)

Because they are supplementary,

$$2x + (5x+75) = 180, \quad x = 15$$
$$z = 2x = 2 \cdot 15 = 30$$

5. (C)

$$5l = 60, \quad l = 12$$
$$5^2 + 12^2 = d^2, \quad 169 = d^2, \quad 13 = d$$

6. (E)

Because of alternate interior angles of parallel lines,

$$3x + 19 = 8x - 1, \quad 5x = 20, \quad x = 4$$

Angle z and 8x-1 are supplementary, so

$$z + (8x-1) = 180, \quad z + 31 = 180, \quad z = 149$$

7. (D)

$$old = bh$$
$$new = (2b)(3h) = 6bh$$
$$new/old = 6/1$$

8. (B)

$$\left(\sqrt{10}\right)^2 + \left(\sqrt{10}\right)^2 = d^2$$
$$10 + 10 = d^2$$
$$d = \sqrt{20} = 2\sqrt{5}$$
$$d + d = 4\sqrt{5}$$

Unit 4.5 Solids

Fortunately there are only three solids that are popular on the standardized tests: the rectangular solid, the cube, and the cylinder.

Rectangular solids

Take a look below at a diagram of a rectangular solid. Think of a sandbox filled with sand. It has a certain width and a certain length, which define the base of the sandbox, and it has a certain height. The box has six faces. Each face is a rectangle with a surface area (SA) equal to the area of its respective rectangle. The total surface area of the rectangular solid is the sum of the surface areas of the faces.

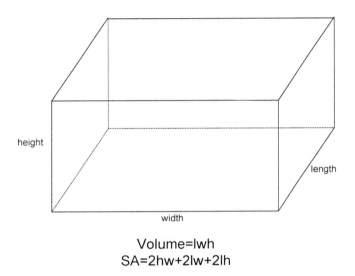

Volume=lwh
SA=2hw+2lw+2lh

Cubes

An important special case of the rectangular solid is the cube. The width, length and height of a cube are all the same, represented by the letter s which stands for side. The volume and surface area of the cube are:

$$Volume = lwh = s \cdot s \cdot s = s^3$$
$$SA = 2hw + 2lw + 2lh = 2s^2 + 2s^2 + 2s^2 = 6s^2$$

Cylinders

When you think of a cylinder, think of a can of soup (or soda or beer, whichever you prefer). The volume of a cylinder is the area of the circular base times the height of the cylinder. The surface area is a bit trickier. It is the sum of the areas of the two circular bases plus the area of the side. To get the area of the side, pretend to have removed both lids of the can. Then cut the can vertically and flatten out the sides. The length of the resulting rectangle will be the circumference of the circular base and the width of the rectangle is the height of the cylinder.

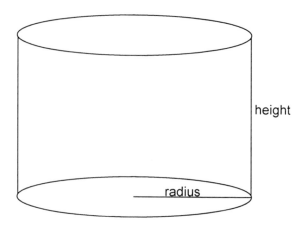

$$Volume = \pi r^2 h$$
$$SA = 2\pi r^2 + 2\pi rh$$

Diagonals of rectangular solids

An important type of question that you might see on a standardized test involves the diagonal of a rectangular solid. When you see a question of this sort, use the Pythagorean Theorem twice. First find the legs of the right triangle formed by the diagonal, then find the diagonal itself.

How many small things fit in a big thing?

Another important type of problem involving solids is the question of how many small solids can fit in a large solid. To find the answer, calculate the volume of the large solid and then divide by the volume of the small solid. For a fun example, find the number of glasses of beer that can be served from a keg (assuming all glasses are the same and all participants are over 21) by dividing the volume of the keg by the volume of the glass.

IMPORTANT NOTE: This same reasoning can be applied to area. How many tiles are needed to cover a floor? Calculate the area of the floor and divide by the area of each tile.

Practice problems on solids (unit 4.5)

1. What is the width of a rectangular solid that has a volume of 375, a length of 5 and a height of 5?

    (A)   5

    (B)   10

    (C)   15

    (D)   25

    (E)   75

2. How many cubes of length 2 would it take to fill a rectangular solid with a volume of 120?

    (A)   15

    (B)   30

    (C)   45

    (D)   60

    (E)   75

3. What is the height of a cylinder whose volume is 18π and whose radius is 3?

    (A)   2

    (B)   4

    (C)   6

    (D)   8

    (E)   10

4. How long is the edge of a cube whose surface area is 150 square units?

    (A)   2

    (B)   5

    (C) $2.5\sqrt{6}$

    (D) $5\sqrt{6}$

    (E)   25

5. What is the surface area of a cylinder with a radius of π and a height of π?

    (A) $2\pi^2$

    (B) $2\pi^3$

    (C) $4\pi^2$

    (D) $4\pi^3$

    (E) $4\pi^4$

6. What is the length of the diagonal of a cube whose side is 3 units?

    (A) $2\sqrt{2}$

    (B) $2\sqrt{3}$

    (C)   3

    (D) $3\sqrt{2}$

    (E) $3\sqrt{3}$

7. A roll of wallpaper is 3 feet wide and 25 feet long. How many rolls will be needed to cover a cube whose edge is 15 feet?

    (A)   3

    (B)   12

    (C)   16

    (D)   18

    (E)   45

8. What is the length of the diagonal of a rectangular solid with a length of 4 and a width of 3 and a height of 6?

    (A)   5

    (B)   6

    (C)   7.8

    (D)   8.5

    (E)   9

9. If 50 guests are expected at a party, and each guest is expected to drink 3 cylindrical glasses of punch, and each glass has a radius of 5cm and a height of 10 cm, what must be the minimum capacity of the punch bowl (in cubic centimeters)?

    (A) 7500π

    (B) 12500π

    (C) 25000π

    (D) 37500π

    (E) 50000π

Solutions to practice problems on solids (unit 4.5)

1. (C)

$$375 = 5 \cdot 5 \cdot h, \quad 15 = h.$$

2. (A)

$$\frac{120}{2^3} = 15$$

3. (A)

$$18\pi = \pi 3^2 h, \quad 2 = h$$

4. (B)

$$SA = 150 = 6s^2$$
$$25 = s^2, \ 5 = s$$

5. (D)

$$SA = 2\pi r^2 + 2\pi r h = 2\pi^3 + 2\pi^3 = 4\pi^3$$

6. (E)

Find the diagonal of the base:
$$3^2 + 3^2 = d^2, \quad 3\sqrt{2} = d$$

Find the diagonal of the cube:
$$3^2 + \left(3\sqrt{2}\right)^2 = d^2, \quad 9 + 18 = d^2, \quad 3\sqrt{3} = d$$

7. (D)

Each roll has $3 \cdot 25 = 75$ square feet.

Cube has $15 \cdot 15 \cdot 6 = 1350$ square feet.

1350/75 = 18 rolls

8. (C)

Find the diagonal of the base:
$$3^2 + 4^2 = d^2, \quad 5 = d$$

Find the diagonal of the cube:
$$6^2 + 5^2 = d^2, \quad 36 + 25 = d^2, \quad \sqrt{61} = d$$

9. (D)

Glasses are $50 \cdot 3 = 150$

Each glass holds $\pi \cdot 5^2 \cdot 10 = 250\pi$

Capacity must be $150 \cdot 250\pi = 37500\pi$

Unit 4.6 Slope, distance and midpoint

Remember how to find the equation of a line? The general equation is

$$y = mx + b,$$

where m is the slope of the line and b is where the line crosses the y-axis (the y-intercept).

The equation of a line can be determined if you know two of the points through which it travels (two points determine a line). We denote the points as (x_1, y_1) and (x_2, y_2). The first step is to find the slope

$$m = \frac{y_2 - y_1}{x_2 - x_1}.$$

Once we have the slope, we can use one of the points to find the equation of the line. For example, suppose we want the equation of the line that passes through (1,2) and (5, 14). We determine the slope to be

$$m = \frac{14 - 2}{5 - 1} = \frac{12}{4} = 3.$$

Next write the partial equation

$$y = 3x + b.$$

Substitute the values (1,2) in the partial equation and solve for b:

$$2 = 3 \cdot 1 + b, \quad -1 = b.$$

So we know that the equation is: $y = 3x - 1$.

Instead of providing two points, some problems will provide the slope of the line and a single point (the slope and a point determine a line). If a problem gives the slope and a point, simply write the partial equation, plug in the point and solve for b. For example, given a slope of -3 and the point (2,3) we write the partial equation

$$y = -3x + b$$

then plug in (2,3) to solve for b

$$3 = -3(2) + b, \quad 9 = b \ .$$

The equation of the line is $y = -3x + 9$.

Rather than provide the slope directly, sometimes the slope is provided indirectly through information about another line. In order to do this type of problem you must remember that:

- The slopes of parallel lines are the same
- The slopes of perpendicular lines are negative reciprocals of each other.

For example, if a line has a slope of 3 then all lines parallel to it must also have a slope of 3; whereas all lines perpendicular to it must have a slope of $-\dfrac{1}{3}$.

There are two special cases of slope to keep in mind:

- The slope of a horizontal line is zero. For example, the line $y = 3$ has a slope of zero.
- The slope of a vertical line is undefined. For example, the slope of the line $x = 3$ is undefined.

Distance formula

The distance formula is something that you should memorize, or if memorizing does not suit you, remember that it is a simple application of the Pythagorean Theorem. You are given two points and need to find the distance between them.

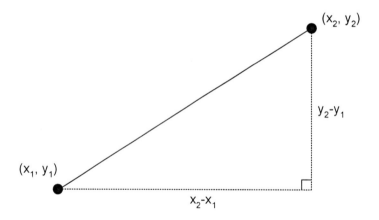

Notice that the vertical distance is just $y_2 - y_1$ and the horizontal distance is just $x_2 - x_1$. Applying the Pythagorean Theorem gives

$$d^2 = \left(y_2 - y_1\right)^2 + \left(x_2 - x_1\right)^2$$
$$d = \sqrt{\left(y_2 - y_1\right)^2 + \left(x_2 - x_1\right)^2}$$

Midpoint formula

The midpoint of the line segment with endpoints $\left(x_1, y_1\right)$ and $\left(x_2, y_2\right)$ has the coordinates:

$$\text{midpoint} = \left(\frac{x_1 + x_2}{2}, \ \frac{y_1 + y_2}{2} \right)$$

Problems on slope, distance and midpoint (unit 4.6)

1. The equation of the line passing through points (5,7) and (-15, 23) is:

(A) $y = -0.8x + 11$

(B) $y = -0.8x + 3$

(C) $y = 0.8x + 12.6$

(D) $y = 0.8x - 0.6$

(E) $y = -1.25x + 13.25$

2. What is the length of the line segment whose endpoints are (5,7) and (-15, 23)?

(A) 18.9

(B) 25.6

(C) 36.0

(D) 656.0

(E) 900.0

3. Find the equation of the line that passes through (1,3) and is perpendicular to the line with the equation $3x - 2y = 5$.

(A) $y = 3x$

(B) $2y = 3x + 3$

(C) $2y = -3x + 9$

(D) $3y = 2x + 4$

(E) $3y = -2x + 11$

4. What is the slope of the line that passes through the points (3, -2) and (3, 2)?

(A) 0.0

(B) 0.4

(C) 0.6

(D) 0.75

(E) Cannot be determined

5. The diameter of a circle has as its endpoints (1,2) and (9, 24). What are the coordinates of the center?

(A) (4, 11)

(B) (4, 13)

(C) (5, 11)

(D) (5, 13)

(E) Cannot be determined

6. If two vertices of rectangle ABCD have coordinates A=(1,2) and B=(5,8), what is the slope of line segment BC?

(A) $-2/3$

(B) $-3/2$

(C) $2/3$

(D) 1

(E) $3/2$

7. What is the equation of the perpendicular bisector of the line segment whose endpoints are (5,7) and (-5,11)?

(A) $y = 0.4x + 9$

(B) $y = -0.4x + 9$

(C) $y = -0.4x - 9$

(D) $y = 2.5x + 9$

(E) $y = 2.5x - 5.5$

8. The line y=3 passes through all of the points below EXCEPT

I. (1, 3)

II. (3,1)

III. (3,3)

(A) I only

(B) II only

(C) III only

(D) I and III only

(E) II and III only

9. If triangle ABC is isosceles and two of its vertices are (0, 0) and (3, 4), which of the following could be its third vertex?

I. (-3, 4)

II. (3, -1)

III. (-3, -4)

(A) I only

(B) II only

(C) III only

(D) I and II only

(E) I, II, and III

Solutions to problems on slope, distance and midpoint (unit 4.6)

1. (A)

$$m = \frac{23-7}{-15-5} = \frac{-4}{5}, \quad y = \frac{-4}{5}x + b$$

at $(5,7)$ $\quad 7 = \frac{-4}{5}(5) + b, \quad 11 = b$

$$y = \frac{-4}{5}x + 11$$

2. (B)

$$d = \sqrt{(23-7)^2 + (-15-5)^2} = \sqrt{256 + 400} = \sqrt{656}$$

3. (E)

$$3x - 2y = 5, \quad -2y = -3x + 5, \quad y = \frac{3}{2}x - 2.5$$

$$y = \frac{-2}{3}x + b \quad \text{at } (1,3) \quad 3 = \frac{-2}{3}(1) + b, \quad \frac{11}{3} = b$$

$$y = \frac{-2}{3}x + \frac{11}{3} \quad \text{or} \quad 3y = -2x + 11$$

4. (E)

This a vertical line with the equation is $x = 3$. Its slope is undefined.

5. (D)

$$\text{center=midpoint} = \left(\frac{1+9}{2}, \frac{2+24}{2}\right) = (5,13)$$

6. (A)

$$\text{slope of AB} = \frac{8-2}{5-1} = \frac{3}{2}$$

$$\text{slope BC} = \frac{-2}{3}$$

7. (D)

$$\text{original slope} = \frac{11-7}{-5-5} = \frac{4}{-10}$$

desired slope = $10/4 = 2.5$

$$\text{midpoint} = \left(\frac{5-5}{2}, \frac{11+7}{2}\right) = (0,9)$$

at $(0, 9)$, equation is $y = 2.5x + 9$

8. (B)

II only.

I. is false because the line passes through (1,3)

II is true because the line does not pass through (3,1)

III is false because the line passes through (3,3)

9. (D)

$(0,0)$ to $(3,4)$ is 5 units, so at least one other side must also be 5 units.

I. is true because $(0,0)$ to $(-3,4)$ is 5 units

II is true because $(3,4)$ to $3,-1)$ is 5 units

III is false even though $(0,0)$ to $(-3,-4)$ is 5 units, because $(3,4)$, $(0,0)$ and $(-3,-4)$ all lie on the straight line with equation $y = \frac{4x}{3}$.

5. ALGEBRA 2

Algebra 2 is a loose collection of topics, and fortunately not all of them are on the SAT®. There are just a few that need to be covered here. The ACT® and SAT® Subject Tests contain much more material from Algebra 2, and those topics are covered in Chapter 7.

5.1 Sequences

A sequence is a collection of terms, denoted as $a_1, a_2, a_3, \ldots$ where a_1 is the first term in the sequence, a_2 is the second term in the sequence, and so on. In general a_n is the n<u>th</u> term in the sequence. There are two kinds of sequences that you need to know.

Arithmetic sequences

The key idea behind an arithmetic sequence is called the common difference, d. If you take any term in an arithmetic sequence and subtract the previous term, the difference is always the same. For example, the sequence

$$-3, 1, 5, 9, \ldots$$

is an arithmetic sequence with a common difference of 4. If you take any term and subtract the previous term, the difference is always 4. In this example, $1 - -3 = 4$ and $5 = 1 = 4$. In general the n<u>th</u> term of an arithmetic sequence is

$$a_n = a_1 + (n-1)d.$$

Geometric sequences

The key idea behind a geometric sequence is called the common ratio, r. If you take any term in the sequence and divide it by the previous term, the ratio is always the same. For example, the sequence

$$-2, 6, -18, 54, \ldots$$

is a geometric sequence with a common ratio of -3. If you take any term and divide by the previous term, the ratio is always -3. In this example, $\dfrac{6}{-2} = -3$ and $\dfrac{-18}{6} = -3$. In general the n<u>th</u> term of a geometric sequence is

$$a_n = a_1 (r)^{n-1}.$$

Problems on sequences (unit 5.1)

The sequence below applies to problems 1 and 2:

$$3, -1, \underline{\quad}, -9$$

1. What is the missing value in this sequence?

    (A)   -6

    (B)   -5

    (C)   -4

    (D)   0

    (E)   4

2. What is the 33rd element of this sequence?

    (A)   -125

    (B)   -128

    (C)   -129

    (D)   -131

    (E)   -132

The sequence below applies to problems 3 and 4:

$$-3, \underline{\quad}, -12, 24$$

3. What is the missing value in this sequence?

    (A)   -6

    (B)   -2

    (C)   0

    (D)   2

    (E)   6

4. What is the 19th element of this sequence?

    (A)   -1,572,864

    (B)   -786,432

    (C)   -262,144

    (D)   786,432

    (E)   -1,572,864

5. A study of fish in an experimental tank found that the number of fish doubles every 5 years. If the number of fish is now 10, how many fish will be in the tank 20 years from now?

    (A)   160

    (B)   320

    (C)   2,621,440

    (D)   5,242,880

    (E)   10,485,760

6. At the end of every year John deposits $1,500 into his savings account. How many years will it take for John's deposits to amount to $19,500?

    (A)   12

    (B)   13

    (C)   14

    (D)   15

    (E)   16

7. The number of carbon particles in the air over a certain city is reduced by 50% every year. If there are 1,499,968 particles now, how many years will it take to reduce the number to 23,437 particles?

    (A)   4

    (B)   5

    (C)   6

    (D)   7

    (E)   8

8. A certain investment pays 7% every year. If $20,000 is invested today, what will that investment be worth 10 years from now?

    (A)   12.600

    (B)   21,400

    (C)   36,769

    (D)   39,343

    (E)   42,097

9. If a car depreciates in value by 20% per year, what percentage of its original value will the car be worth when it reaches its third birthday?

    (A)   0.8%

    (B)   4%

    (C)   41%

    (D)   51%

    (E)   64%

Solutions to problems on sequences (unit 5.1)

1. (B)

The common difference is -4.

3, -1, -5, -9

2. (A)

$$a_n = a_1 + (n-1)d = 3 + (33-1)(-4)$$
$$= 3 + 32(-4) = -125$$

3. (E)

The common ratio is -2

-3, 6, -12, 24

4. (B)

$$a_n = a_1 (r)^{n-1} = (-3)(-2)^{19-1}$$
$$= -3(-2)^{18} = -786,432$$

5. (A)

First realize that $a_1 = 10$. The "trick" is to realize that the problem is asking for a_5 because four periods of 5 years each will have elapsed after 20 years.

$$a_5 = 10(2)^4 = 160$$

6. (B)

$$a_n = a_1 + (n-1)d$$
$$19,500 = 1500 + (n-1)1,500$$
$$19,500 = 1500n, \ 13 = n$$

It will take 13 years to save 19,500.

7. (B)

$$a_n = a_1 \left(\frac{1}{2}\right)^n$$

$$23,437 = 1,499,968 \left(\frac{1}{2}\right)^n$$

$$\frac{1}{64} = \left(\frac{1}{2}\right)^n, \ n = 6$$

It will take five more years, because the first term is the current value.

8. (D)

$$a_{11} = 20,000(1.07)^{10} = 39,343.03$$

9. (D)

$$(0.8)^3 = 0.512 = 51\%$$

Or substitute values using a starting value of 100:

100, 80, 64, 51.2

5.2 Absolute value

Absolute value problems can be tricky, especially inequalities. But there is a technique that will help you solve them easily. First, we begin with a definition of absolute value:

$$\text{When } x \geq 0, \ |x| = x. \quad \text{When } x < 0, \ |x| = -x.$$

Note that the absolute value of a number is always positive. If x is positive in the first place, leave the sign alone. If x is negative, change the sign to positive.

Equalities

The simplest absolute value problems to solve are equalities, where the absolute value of some expression is set equal to a non-negative constant. The figure below illustrates how two equations are spawned and where the solutions fall on a number line.

For example, the solutions to the equation $|x| = 10$ are $x = 10$ or $x = -10$. When the expression inside the absolute value is more complicated, the process is the same. Simply take the expression and spawn two equations, equating that expression to the positive or negative value of the constant.

Less than (or less than or equal to)

Absolute values can also appear in inequalities. The figure below shows how an inequality is handled when an absolute value is in a less than (or less than or equal to) a quantity. Note that the solution is an interval. The figure illustrates how two equations are spawned to find the interval, along with the solution on a number line.

For example, the solution to the equation $|x| < 10$ is the interval $-10 < x < 10$. When the expression inside the absolute value is more complicated, the process is the same. Take the expression and spawn two inequalities in order to find the appropriate interval.

Greater than (or greater than or equal to)

The figure below shows how an inequality is handled when an absolute value is in a greater than (or greater than or equal to) a quantity. Note that the solution is two intervals. The figure illustrates how two equations are spawned to find the intervals, along with how the solution looks on a number line.

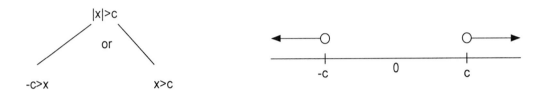

For example, the solution to the equation $|x| > 10$ are the intervals $-10 > x$ or $x > 10$. When the expression inside the absolute value is more complicated, the process is the same. Take the expression and spawn two inequalities in order to find the appropriate intervals for the solution.

Problems on absolute value (unit 5.2)

1. The solutions to the equation $5|x-2|=20$ are:

    (A)   x=-2 and x=2

    (B)   x=-2 and x=4

    (C)   x=-2 and x=6

    (D)   x=-4 and x=2

    (E)   x=-4 and x=4

2. The solution to the inequality $2|x+5|<8$ is:

    (A) $-9<x<-1$

    (B) $-9<x<1$

    (C) $-1<x<1$

    (D) $-1<x<9$

    (E) $1<x<9$

3. The solution to the inequality $5|x-3|-5\leq0$ is:

    (A) $-4\leq x\leq-2$

    (B) $-4\leq x\leq2$

    (C) $-2\leq x\leq2$

    (D) $-2\leq x\leq4$

    (E) $2\leq x\leq4$

4. The solution to the inequality $3|x+8|\geq6$ is:

    (A) $-10\geq x\ or\ x\geq-6$

    (B) $-10\geq x\ or\ x\geq6$

    (C) $-6\geq x\ or\ x\geq6$

    (D) $-6\geq x\ or\ x\geq10$

    (E) $6\geq x\ or\ x\geq10$

5. The solution to the inequality $|2x-5|>9$ is:

    (A) $-7\geq x\ or\ x\geq-2$

    (B) $-7\geq x\ or\ x\geq2$

    (C) $-7\geq x\ or\ x\geq7$

    (D) $-2\geq x\ or\ x\geq7$

    (E) $-2\geq x\ or\ x\geq-7$

6. The solution to the inequality $5|3x|>-15$ is:

    (A) $-1\geq x\ or\ x\geq1$

    (B) $1\geq x\ or\ x\geq-1$

    (C) $x\geq0$

    (D)   All values of x

    (E)   No solution

7. The solution to the inequality $|2x|<-2$ is:

    (A) $-2<x<1$

    (B) $-1<x<1$

    (C) $1<x<2$

    (D)   All values of x

    (E)   No solution

8. In order to be admitted to a ride in an amusement park, children must be older than three and younger than seven. Which of the inequalities below represents the admission requirement?

    I. $3<x<7$

    II. $|x-3|<7$

    III. $|x-5|<2$

    (A)   I only

    (B)   II only

    (C)   III only

    (D)   I and II only

    (E)   I and III only

Solutions to problems on absolute value (unit 5.2)

1. (C)

Isolate the absolute value: $|x-2| = \dfrac{20}{5} = 4$

First equation: $x-2=-4, \quad x=-2$

Second equation: $x-2=4, \quad x=6$

2. (A)

Isolate the absolute value: $|x+5| < \dfrac{8}{2} = 4$

First equation: $-4 < x+5, \quad -9 < x$

Second equation: $x+5 < 4, \quad x < -1$

Solution: $-9 < x$ and $x < -1$

3. (E)

Isolate the absolute value: $5|x-3| \le 5, \quad |x-3| \le 1$

First equation: $-1 \le x-3, \quad 2 \le x$

Second equation: $x-3 \le 1, \quad x \le 4$

4. (A)

Isolate the absolute value: $3|x+8| \ge 6, \quad |x+8| \ge 2$

First equation: $-2 \ge x+8, \quad -10 \ge x$

Second equation: $x+8 \ge 2, \quad x \ge -6$

5. (D)

First equation: $-9 > 2x-5, \quad -4 > 2x, \quad -2 > x$

Second equation: $2x-5 > 9, \quad 2x > 14, \quad x > 7$

6. (D)

Isolate the absolute value: $5|3x| > -15, \quad |3x| > -3$

This inequality is true for all values of x. The absolute value of any quantity is always non-negative.

7. (E)

This inequality is never true for any value of x. The absolute value of any quantity is always non-negative.

8. (E)

I. obviously is correct.

II is incorrect:

$|x-3| < 7$

$-7 < x-3 < 7$

$-7+3 < x < 7+3$

$-4 < x < 10$

III is correct

$|x-5| < 2$

$-2 < x-5 < 2$

$-2+5 < x < 2+5$

$3 < x < 7$

5.3 Exponents and radicals

The use of exponents is an important part of all of the standardized tests. We begin first with the laws of exponents – these **must** be memorized!

$$x^a \cdot x^b = x^{a+b} \qquad\qquad\qquad x^0 = 1$$

$$\frac{x^a}{x^b} = x^{a-b} \qquad\qquad\qquad (xy)^a = x^a y^a$$

$$\left(x^a\right)^b = x^{ab} \qquad\qquad\qquad \left(\frac{x}{y}\right)^a = \frac{x^a}{y^a}$$

$$x^{-a} = \frac{1}{x^a} \qquad\qquad\qquad \frac{1}{x^{-a}} = x^a$$

In addition to memorizing these laws, of course you must know how to apply them. There are several problems at the end of this unit to test that ability.

Equations involving exponents

Sometimes you will see equations involving exponents. The technique to use here is to convert both sides of the equation to the same base. Once the bases are the same, you can equate the exponents. For example:

$$4^x = 128 \qquad\qquad \text{original equation}$$

$$\left(2^2\right)^x = 2^7 \qquad\qquad \text{convert both sides to use base 2}$$

$$2^{2x} = 2^7 \qquad\qquad \text{simplify the left-hand side using a law of exponents}$$

$$2x = 7, \quad x = 3.5 \qquad\qquad \text{equate exponents; solve for x.}$$

Radicals and exponents

To some extent, radicals were covered in unit 3.2. However, there we restricted our attention to a particular type of radical, the square root.

Perhaps the best way to deal with square roots and other radicals is to change them into fractional exponents using the law:

$$\sqrt[n]{x} = x^{\frac{1}{n}}.$$

Once the radical is written as a fractional exponent, it can be manipulated using the laws of exponents given above. Also, using fractional exponents permits us to solve problems easily on the calculator

Calculator tip: What if you were asked to find the 7th root of 16,384. That is an easy job for the calculator. Just 16384, ^, (, 1/7,), enter. The calculator returns 4. It turns out that $4^7 = 16,384$, not exactly common knowledge.

Fractional exponents also can be used to simplify combinations of powers and roots, as show in the example below:

$$\sqrt[6]{x^3} = \left(x^3\right)^{\frac{1}{6}} = x^{\frac{3}{6}} = \sqrt{x} \quad \text{and} \quad \left(\sqrt[6]{x}\right)^3 = \left(x^{\frac{1}{6}}\right)^3 = x^{\frac{3}{6}} = \sqrt{x}$$

Problems on exponents and radicals (unit 5.3)

1. The expression $x^3 \cdot x^9$ simplifies to

   (A) x^6

   (B) x^{12}

   (C) x^{27}

   (D) x^3

   (E) x^{-6}

2. The expression $\dfrac{a^8}{a^{11}}$ simplifies to

   (A) a^3

   (B) a^{19}

   (C) a^{88}

   (D) a^{-2}

   (E) a^{-3}

3. The expression $\sqrt{3^{30} x^4}$ simplifies to:

   (A) $(3x)^{68}$

   (B) $(3x)^{34}$

   (C) $(3x)^{17}$

   (D) $3^{15} x^2$

   (E) $3^{60} x^8$

4. The expression $\dfrac{\left(\sqrt[3]{y}\right)^5}{y^2}$ simplifies to:

   (A) $y^{\frac{11}{3}}$

   (B) y^3

   (C) $y^{\frac{1}{3}}$

   (D) $y^{-\frac{1}{3}}$

   (E) $y^{-\frac{11}{3}}$

5. Solve $9^{x+1} = \sqrt{3^{x+2}}$ for x:

   (A) $x = -\dfrac{2}{3}$

   (B) $x = 0$

   (C) $x = \dfrac{2}{3}$

   (D) $x = 1$

   (E)   No solution

6. Solve $81^x = \sqrt[3]{27^2}$ for x:

   (A) $x = -\dfrac{4}{3}$

   (B) $x = \dfrac{1}{2}$

   (C) $x = \dfrac{3}{4}$

   (D) $x = \dfrac{4}{3}$

   (E) $x = 2$

7. Simplify $\dfrac{x^2 y}{z} \div \dfrac{x}{zy}$

   (A) xy^2

   (B) $x^2 y^2$

   (C) $x^3 y^2$

   (D) $x^3 y^3$

   (E) $\dfrac{x^3}{z^2}$

8. Simplify $\dfrac{x^{12} y^{-2} z^5}{x^3 y^4 z^{-3}}$

   (A) $x^8 y^{-6} z^8$

   (B) $x^8 y^2 z^8$

   (C) $x^9 y^6 z^8$

   (D) $x^9 y^{-6} z^8$

   (E) $x^{15} y^2 z^2$

Solutions to problems on exponents and radicals (unit 5.3)

8. (D)

$$\frac{x^{12}y^{-2}z^5}{x^3y^4z^{-3}} = \left(\frac{x^{12}}{x^3}\right)\left(\frac{y^{-2}}{y^2}\right)\left(\frac{z^5}{z^{-3}}\right) = x^{12-3}y^{-2-4}z^{5--3} = x^9y^{-6}z^8$$

1. (B)

$$x^3 \cdot x^9 = x^{3+9} = x^{12}$$

2. (E)

$$\frac{a^8}{a^{11}} = a^{8-11} = a^{-3}$$

3. (D)

$$\sqrt{3^{30}x^4} = \left(3^{30}\right)^{\frac{1}{2}}\left(x^4\right)^{\frac{1}{2}} = 3^{15}x^2$$

4. (D)

$$\frac{\left(\sqrt[3]{y}\right)^5}{y^2} = \frac{y^{\frac{5}{3}}}{y^2} = y^{\frac{5}{3}-\frac{6}{3}} = y^{-\frac{1}{3}}$$

5. (A)

$$9^{x+1} = \sqrt{3^{x+2}}, \quad \left(3^2\right)^{x+1} = \left(3^{x+2}\right)^{\frac{1}{2}},$$

$$2(x+1) = \frac{1}{2}(x+2), \quad \frac{3x}{2} = -1, \quad x = \frac{-2}{3}$$

6. (B)

$$81^x = \sqrt[3]{27^2}, \quad \left(3^4\right)^x = \left(\left(3^3\right)^2\right)^{\frac{1}{3}}, \quad 3^{4x} = 3^2, \quad x = 0.5$$

7. (A)

$$\frac{x^2y}{z} \div \frac{x}{zy} = \left(\frac{x^2y}{z}\right)\left(\frac{zy}{x}\right) = xy^2$$

5.4 Parabolas

A parabola is a graph of a second-degree equation (also called a quadratic equation), where y is set equal to terms involving x^2. You need to be able to recognize quadratic equations, graph them, understand the symmetry of the parabola, and find the x-intercepts (also called zeros or roots of the equation).

When solving problems involving quadratic equations, do not forget that you can use your graphing calculator. The calculator can graph the parabola, find the vertex of the parabola, and find the zeros (x-intercepts) of the parabola.

Equations in Standard Form

A quadratic equation in standard form is $y = ax^2 + bx + c$.

The graph of a quadratic equation is a parabola. The parabola opens up if $a > 0$ and the parabola opens down if $a < 0$. The parabola is symmetric about its axis of symmetry, which is an invisible line running vertically through the vertex. The equation of the axis of symmetry is $x = \dfrac{-b}{2a}$. This is illustrated in the figures below.

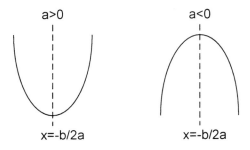

The x-coordinate of the vertex is found by using the formula: $x = \dfrac{-b}{2a}$. The y-coordinate of the vertex is found by plugging the x-coordinate into the equation.

Calculator tip: After entering the equation into the equation editor (y= editor), use the 2nd-calc-minimum (parabola opens up) or 2nd-calc-maximum function (parabola opens down) to find the vertex. To find the roots (x-intercepts), use the 2nd-calc-zero function. To find the y-intercept, use the 2nd-calc-value function, plugging in zero for x.

Equations in Vertex Form

If the goal is to graph the parabola by hand, it is easier to do so when the equation is written in vertex form rather than standard form. The vertex form of the equation is

$$y = a(x - h)^2 + k$$

The vertex is (h,k) and the axis of symmetry is the equation $x=h$. This is illustrated in the figures below.

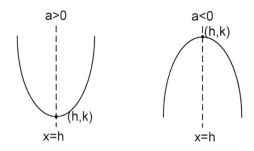

Finding the roots (zeros) of the equation

The zeros or roots of the equation are obtained by finding the values of x for which $y = f(x) = 0$. The roots are the x-coordinates of the x-intercepts of the parabola.

Although there are several methods to find the roots, the method used in this unit is to factor the equation (for more on factoring, see unit 3.5) and then set each factor to zero.

Once factored, the equation of the parabola may be written in intercept form:

$$y = a(x - r_1)(x - r_2),$$

where $(r_1, 0)$ and $(r_2, 0)$ are the roots (x-intercepts) of the parabola. The x-coordinate of the vertex lies halfway between the roots.

By graphing the equation, you can see what kinds of roots exist. If the parabola never crosses the x-axis then the equation has two imaginary roots. If the parabola crosses the x-axis at two points, then the equation has two real roots. If the vertex of the parabola lies on the x-axis then the equation has one real root (sometimes called a double root).

Problems on parabolas (unit 5.4)

TRY TO NOT USE YOUR CALCULATOR TO SOLVE ANY OF THESE PROBLEMS!

1. Which of these parabolas opens up?

    I. $y = 2x^2 - 5x - 3$

    II. $y = -2x^2 - 5x - 5$

    III. $y = 2x^2 + 5x + 3$

(A)   I only

(B)   II only

(C)   III only

(D)   I and II only

(E)   I and III only

2. Which of these parabolas **must** have two real roots?

    I. $y = 2x^2 - 5x - 3$

    II. $y = -2x^2 - 5x - 5$

    III. $y = 2x^2 + 5x - 1$

(A)   I only

(B)   II only

(C)   III only

(D)   I and II only

(E)   I and III only

3. If a parabola crosses the x-axis at (-5,0) and (3,0) then its axis of symmetry must be:

(A) $x = -5$

(B) $x = -3$

(C) $x = -1$

(D) $x = 1$

(E) $x = 5$

4. The vertex of the parabola $y + 16 = (x - 1)^2$ is located at:

(A)   (-1, -16)

(B)   (-1, 16)

(C)   (1, -16)

(D)   (1, 16)

(E)   (16, 1)

NOTE: Questions 5 and 6 pertain to the parabola: $y = 2x^2 + x - 6$.

5. Find the axis of symmetry:

(A) $x = -0.5$

(B) $x = -0.25$

(C) $x = 0$

(D) $x = 0.25$

(E) $x = 0.5$

6. Find the roots (zeros):

(A) $x = -1, 3$

(B) $x = 1, 3$

(C) $x = 1.5, -2$

(D) $x = -1.5, 2$

(E)   Cannot be determined

7. In the diagram below, the coordinates of (a, b) are:

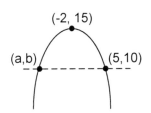

(A)   (-5, 10)

(B)   (-5, 15)

(C)   (-9, 10)

(D)   (-9, 15)

(E)   (10, 5)

8. If the roots of a parabola are -3 and 5, then the equation of the parabola is:

(A) $y = x^2 - 2x - 15$

(B) $y = x^2 + 2x - 15$

(C) $y = x^2 + 8x + 15$

(D) $y = x^2 - 8x + 15$

(E)   Cannot be determined

Solutions to problems on parabolas (unit 5.4)

1 (E)

A parabola opens up if the coefficient of the x^2 term is positive. That coefficient is positive in the first and third equations.

2. (E)

Equation I has a y-intercept of -3 and it opens up. Therefore it must cross the x-axis in two places. Equation II has a y-intercept of -5 and opens down. Therefore it can never cross the x-axis. Equation III has a y-intercept of -1 and opens up. Like Equation I, it must cross the x-axis at two points.

3. (C)

The x-intercepts are equidistant from the axis of symmetry. The value of -1 is 4 units from -5 and 4 units from +3.

4. (C)

Rewrite the equation:

$$y + 16 = (x - 1)^2$$
$$y = (x - 1)^2 - 16$$
$$y = (x - h)^2 + k$$

The vertex is at (h, k) = (1, -16).

5. (B)

The axis of symmetry is

$$x = \frac{-b}{2a} = \frac{-1}{2(2)} = -.25.$$

6. (C)

Factor $y = 2x^2 + x - 6 = (2x - 3)(x + 2)$. To find the roots, set each factor to zero.

$$2x - 3 = 0, \ x = 1.5$$
$$x + 2 = 0, \ x = -2$$

Note that the roots are equidistant from the axis of symmetry found in the previous problem to be -0.25.

7. (C)

This problem uses the symmetry of the parabola. The axis of symmetry is x=-2. If the point (5, 10) is 7 units from the axis of symmetry then the point (a, 10) must also be 7 units from the axis of symmetry, or (-9, 10).

8. (A)

$$(x + 3)(x - 5) = 0$$
$$x^2 + 3x - 5x - 15 = 0$$
$$x^2 - 2x - 15 = 0$$

5.5 Translations and reflections

Translations and reflections are simply movements of the graph of $y = f(x)$. A translation moves the graph horizontally and/or vertically. A reflection causes the graph to flip around an axis. Translations and reflections can take place separately or in combination.

Translations

The table below shows how the graph of $y = f(x)$ is moved vertically and/or horizontally.

Original function $y = f(x)$	Translated function
Move up k units	$y = f(x) + k$
Move down k units	$y = f(x) - k$
Move left h units	$y = f(x + h)$
Move right h units	$y = f(x - h)$

Each movement can take place by itself or in combination with others. For example, to move a graph up k units and right h units you would use $y = f(x - h) + k$.

Reflections

Only the two reflections below are needed:

Original function $y = f(x)$	Reflected function
About the x-axis	$y = -f(x)$
About the y-axis	$y = f(-x)$

Combinations

Translations and reflections are sometimes combined. For example, to move a graph up k units and reflect it about the x-axis would be

$$y = f(x) + k \qquad \text{first translate up k units}$$

$$y = -\big(f(x) + k\big) = -f(x) - k \qquad \text{reflect about x-axis}$$

Problems on translations and reflections (unit 5.5)

1. If the parabola $y = 2x^2 - 3x + 1$ is the result of moving another parabola down 1 unit, what is the equation of the original parabola?

    (A) $y = 2(x+1)^2 - 3(x+1) + 1$

    (B) $y = 3x^2 - 2x$

    (C) $y = 2x^2 - 3x + 2$

    (D) $y = 2x^2 - 3x$

    (E) $y = x^2 - 4x$

2. The equation of the parabola $y = 2x^2 - 3x + 1$

after it has been reflected about the y-axis and then moved up 1 unit is:

    (A) $y = 2(x+1)^2 - 3(x+1) + 1$

    (B) $y = 3x^2 - 2x$

    (C) $y = 2x^2 + 3x + 2$

    (D) $y = 2x^2 - 3x$

    (E) $y = x^2 - 4x$

3. The equation of the parabola $y = 2x^2 - 3x + 1$

after it has been moved right 1 unit and then reflected about the x-axis is:

    (A) $y = 2x^2 - x$

    (B) $y = 2x^2 + x$

    (C) $y = -2x^2 + x$

    (D) $y = -2x^2 + 3x - 1$

    (E) $y = -2x^2 + 7x - 6$

4. The intersection of the line $y = 5x + 3$ and its reflection about the x-axis occurs at the point:

    (A)   (-0.6, 0)

    (B)   (0.6, 0)

    (C)   (0, 0)

    (D)   (0, 3))

    (E)   (3,0)

5. The intersection of the line $y = 5x + 3$ and its reflection about the y-axis occurs at the point:

    (A)   (-0.6, 0)

    (B)   (0.6, 0)

    (C)   (0, 0)

    (D)   (0, 3))

    (E)   (3,0)

6. When the following graph

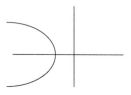

is reflected over the y-axis and then is translated left, it becomes

A.

B.

C.

D.
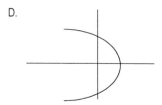

E. None of the above.

Solutions to problems on translations and reflections (unit 5.5)

6. (C)

Graph A is not reflected and shifted left. Graph B is reflected but not shifted. Graph D is not reflected and shifted right.

1. (C)

$$f(x) = 2x^2 - 3x + 2$$
$$f(x) - 1 = 2x^2 - 3x + 1$$

2. (C)

$$f(x) = 2x^2 - 3x + 1$$
$$f(-x) = 2(-x)^2 - 3(-x) + 1$$
$$f(-x) = 2x^2 + 3x + 1$$
$$f(-x) + 1 = 2x^2 + 3x + 2$$

3. (E)

$$f(x) = 2x^2 - 3x + 1$$
$$f(x-1) = 2(x-1)^2 - 3(x-1) + 1$$
$$f(x-1) = 2(x^2 - 2x + 1) - 3(x-1) + 1$$
$$f(x-1) = 2x^2 - 7x + 6$$
$$-f(x-1) = -2x^2 + 7x - 6$$

4. (A)

$$f(x) = 5x + 3$$
$$-f(x) = -5x - 3$$
$$5x + 3 = -5x - 3$$
$$10x = -6, \ x = -0.6$$
$$f(-0.6) = 0$$
$$(-0.6, 0)$$

5. (D)

$$f(x) = 5x + 3$$
$$f(-x) = -5x + 3$$
$$5x + 3 = -5x + 3$$
$$10x = 0, \ x = 0$$
$$f(0) = 3$$
$$(0, 3)$$

5.6 Functions

A function defines a relation between every possible value that can be accepted by the function (this set is called the domain) to every possible value that can be produced by the function (this set is called the range). For example the function $f(x) = x + 3$ establishes a relationship between every domain value and its corresponding range value. The relationship is simple: take the domain value and add three in order to obtain the range value. A useful graphic representation of a function is shown below.

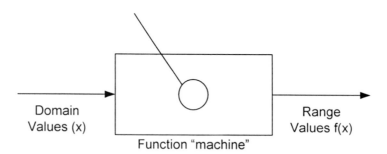

Imagine that you feed a value from the domain (the x-value) into the function machine. Next, turn the crank on the machine. Then out comes the corresponding value from the range (the y-value or the f(x) value). For a particular range value that is placed into it, a function can only crank out one range value (otherwise it is not a function). When functions are graphed, they pass the "vertical line test," meaning that if a vertical line is drawn anywhere on the graph, it passes through the graph of the function at one and only one point.

Funny functions

It is way too easy to ask you to evaluate a traditional function like $f(x) = x^2 + 1$. Instead you are likely to be asked to evaluate functions that use non-mathematical symbols. For example, the # function could be defined as:

$$\# x = \frac{x^2}{2}$$

Then you would be expected to evaluate the # function for various values or expressions in the domain. For example:

$$\# 3 = \frac{3^2}{2} = \frac{9}{2} = 4.5$$

$$\# a = \frac{a^2}{2}$$

$$\#(a-2) = \frac{(a-2)^2}{2}$$

Standardized tests may use funny functions with two arguments, such as

$$x \; \psi \; y = 5x - 3y$$

Then you would be expected to evaluate the ψ function for various expressions, such as:

$$2 \; \psi \; 3 = 5(2) - 3(3) = 10 - 9 = 1$$

$$3 \; \psi \; a = 5(3) - 3a = 15 - 3a$$

$$a \; \psi \; (a-1) = 5a - 3(a-1) = 5a - 3a + 3 = 2a + 3$$

Graphs of functions

It is important to be able to navigate graphs of functions. A general graph of a function is

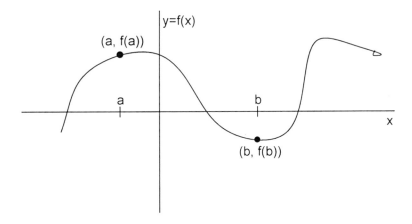

Notice that the domain is the horizontal axis (x-axis) and the range is the vertical axis (y-axis). The function maps each value in the domain to a single value in the range (passes the vertical line test). Any point that lies on the graph must satisfy the function rule. For example if the function were $f(x) = x^2 + 1$ then any arbitrary point on the graph of that function would have the coordinates of $(a, a^2 + 1)$.

Compound funny functions

Sometimes functions are combined together to form compound functions. An example is shown below. Suppose we continue with the # function as before and also introduce the & function as follows:

$$\#x = \frac{x^2}{2} \qquad \&x = x + 2.$$

Then the compound functions are defined as follows:

$$\#(\&x) = \#(x+2) = \frac{(x+2)^2}{2} \quad \text{and} \quad \&(\#x) = \&\left(\frac{x^2}{2}\right) = \frac{x^2}{2} + 2 = \frac{x^2 + 4}{2}.$$

A function can also be compounded on itself as in:

$$\#(\#x) = \#\left(\frac{x^2}{2}\right) = \frac{\left(\frac{x^2}{2}\right)^2}{2} = \frac{x^4}{8} \quad \text{and} \quad \&(\&x) = \&(x+2) = (x+2) + 2 = x + 4.$$

Problems on functions (unit 5.6)

NOTE: For problems 1-3, the function $⇑$ is defined as $⇑ x = 2x^2 - 3$.

1. The value of $⇑ 5$ is:

    (A)   0

    (B)   7

    (C)   10

    (D)   47

    (E)   50

2. The value of $⇑\left(⇑ 2\right)$ is:

    (A)   0

    (B)   7

    (C)   10

    (D)   47

    (E)   50

3. Solve for x when $⇑ x = ⇑\left(x+1\right)$

    (A)   -0.5

    (B)   0

    (C)   0.5

    (D)   1

    (E)   No solution

NOTE: For problems 4 and 5, the function $x\Psi y$ is defined as $x\Psi y = x^2 - xy$.

4. Find the expression for $\left(a+1\right)\Psi\left(a-1\right)$

    (A)   0

    (B)   2

    (C)   2a

    (D)   a+1

    (E)   2a+2

5. If $x\Psi 2 = x\Psi 3y$, then y is equal to:

    (A)   -2/3

    (B)   -3/2

    (C)   2/3

    (D)   3/2

    (E)   2

NOTE: The diagram below is to be used for problems 6-8.

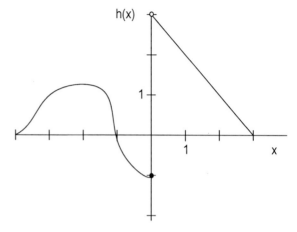

6. The domain of h is:

    (A)   [-4, 1]

    (B)   [-4,0] U (0,3]

    (C)   [-4,3]

    (D)   [-1,1]

    (E)   [-1,3]

7. The value of h(0) is:

    (A)   -1

    (B)   0

    (C)   1

    (D)   2

    (E)   3

8. For what value of k does h(k) =1?

    I. k= -2

    II. k= 2

    III. k= 3

    (A)   I only

    (B)   II only

    (C)   III only

    (D)   I and II only

    (E)   I and III only

Solutions to problems on functions (unit 5.6)

1. (D)

$$\Uparrow 5 = 2(5)^2 - 3 = 50 - 3 = 47$$

2. (D)

$$\Uparrow (\Uparrow 2) = \Uparrow (2 \cdot 2^2 - 3) = \Uparrow (5)$$

$$\Uparrow 5 = 2(5)^2 - 3 = 50 - 3 = 47$$

3. (A)

$$\Uparrow x = 2x^2 - 3$$

$$\Uparrow (x+1) = 2(x+1)^2 - 3 = 2(x^2 + 2x + 1) - 3$$

$$\Uparrow (x+1) = 2x^2 + 4x - 1$$

$$2x^2 - 3 = 2x^2 + 4x - 1$$

$$-2 = 4x, \quad -0.5 = x$$

4. (E)

$$(a+1)\Psi(a-1) = (a+1)^2 - (a+1)(a-1)$$

$$= (a^2 + 2a + 1) - (a^2 - 1) = 2a + 2$$

5. (C)

$$x\Psi 2 = x\Psi 3y$$

$$x^2 - 2x = x^2 - 3xy$$

$$y = \frac{-2x}{-3x} = 2/3$$

6. (C)

The function is defined for all x within [-4,3].

7. (A)

  h(0)=-1

8. (D)

Draw the horizontal line y=1. It intersects the function at -2 and 2.

6. MISCELLANEOUS TOPICS

This chapter contains some topics from probability and statistics, along with word problems.

6.1 Mean, median and mode

This is usually covered in a statistics course or maybe some early math course. Some students claim to have never seen this, so take a look before skipping to another section.

Mean

Standardized tests contain the phrase "average (arithmetic mean)." Don't let this confuse you. The phrase refers to the good ole average, something you learned in grade school. It is the sum over the count. Although details are in unit 2.7, a simple example is:

> "Find the average (arithmetic mean) of 2, 3, -1 and 0." The sum is 4 and the count is 4. So the average is 4/4 or 1.

Median

The value of the median depends on whether the count is odd or even. To find the median, list the numbers from smallest to largest. If the count is odd, the median is the middle number. If the count is even, the median is the mean of the middle two numbers.

> "Find the median of 5, -1, 6, 2, and 0." First, list the numbers from smallest to largest. The ordered list is: -1, 0, 2, 5 and 6. Next we notice that the count is odd (the count is 5). The median is the middle number, 2.

> "Find the median of 5, -1, 6, -3, 2, and 0." First, list the numbers from smallest to largest. The ordered list is: -3, -1, 0, 2, 5 and 6. Next we notice that the count is even (the count is 6). The median is mean of the middle two numbers, which is $\dfrac{0+2}{2} = 1$.

Mode

The mode is the number that appears most frequently. Remember that there can be more than one mode (this is called multi-modal). To find the mode(s), list the numbers from smallest to largest. The mode(s) are the numbers that appear most frequently.

> "Find the mode of 5, -1, 6, -3, 2, 0, and 6." First, list the numbers from smallest to largest. The ordered list is -3, -1, 0, 2, 5, 6, 6. The mode is 6 because it is the only number that appears twice.

> "Find the mode of 2, -1, 6, -3, 2, 0, and 6." First, list the numbers from smallest to largest. The ordered list is -3, -1, 0, 2, 2, 6, 6. The modes are 2 and 6 because they are the only numbers that appear twice.

Problems on mean, median and mode (unit 6.1)

1. For the following numbers: 5, -3, 0, -1, 6, -3, 1 the sum of the median and the mode is:

(A)   -3

(B)   -2

(C)   -1

(D)   0

(E)   1

NOTE: The following numbers are to be used to answer questions 2 and 3:

$$-1, -1, 5, 6, -4, 5$$

2. The difference between the median and the mean is:

(A)   -2

(B)   -1

(C)   1/3

(D)   1

(E)   2

3. The number of modes is:

(A)   0

(B)   1

(C)   2

(D)   3

(E)   4

4. The expression for the average (arithmetic mean) of x, x+2, and 4x+10 is:

(A) $x+2$

(B) $2x+4$

(C) $3x+6$

(D) $4x+8$

(E) $6x+12$

5. If the average (arithmetic mean) of x and 3x is equal to 100, what is the value of x?

(A)   25

(B)   50

(C)   75

(D)   100

(E)   200

NOTE: The table below is to be used to answer questions 6-8.

Mr. Corn asked the 15 students in his math class to report how many television sets are in their homes. The results are tabulated below.

Number of television sets per home	Number of homes
0	1
1	2
2	3
3	4
4	5

6. The average number of televisions per home is:

(A)   2/3

(B)   4/3

(C)   6/3

(D)   8/3

(E)   10/3

7. The median number of televisions per home is:

(A)   2

(B)   2.5

(C)   3

(D)   3.5

(E)   4

8. Suppose there are 16 students in the class, but one was absent that day. When he returns to class he reports the number of television sets in his home. The new median would be:

(A)   2

(B)   2.5

(C)   3

(D)   3.5

(E)   Cannot be determined

Solutions to problems on mean, median and mode (unit 6.1)

1. (A)

The ordered numbers are: -3, -3, -1, 0, 1, 5, 6. The median is 0 because it is the middle number. The mode is -3 because it appears twice. The sum of the median and mode is 0+-3=-3.

2. (C)

The ordered numbers are -4, -1, -1, 5, 5, 6. The median is the mean of -1 and 5. So the median is $\frac{-1+5}{2} = \frac{4}{2} = 2$. The sum is 10 and the count is 6. The mean is 10/6=5/3. The difference between the median and the mean is $2 - \frac{5}{3} = \frac{6}{3} - \frac{5}{3} = \frac{1}{3}$.

3. (C)

The values -1 and 5 each appear twice and no value appears three times, so they are both modes.

4. (B)

$$\frac{\text{sum}}{\text{count}} = \frac{x+(x+2)+(4x+10)}{3} = \frac{6x+12}{3} = 2x+4$$

5. (B)

$$\frac{x+3x}{2} = 100, \quad x+3x = 200, \quad 4x = 200, \quad x = 50$$

6. (D)

$$sum = 0(1)+1(2)+2(3)+3(4)+4(5)$$
$$sum = 0+2+6+12+20 = 40$$
$$avg = sum/count = 40/15 = 8/3$$

7. (C)

The ordered numbers are 0,1,1,2,2,2,3,3,3,3,4,4,4,4,4. The middle number is 3.

8. (C)

The median remains unchanged regardless of how many televisions the sixteenth student has in his home.

If he had 4 or more television sets, the new data would be:

0,1,1,2,2,2,3,3,3,3,4,4,4,4,4,?

If he had less than 3 television sets, the new data would be

0,1,1,2,2,2,?,3,3,3,3,4,4,4,4,4

In either case, the median remains three because the middle two numbers are three.

6.2 Sets and probability

These are two more topics that are usually covered in statistics courses.

Sets

A set is a collection of numbers (or other objects). Each number in the set is called a member or element of the set. A set may be finite or infinite. For example:

$$A = \{1,2,3\} \quad \text{and} \quad B = \{1,3,5,...\}.$$

Set A is finite with three members. Set B contains the positive odd integers. It is infinite.

There are two operations to be done with sets: union and intersection. The union of two sets, denoted by $\cup$, is a collection of elements that belong to either of the parent sets. The intersection of two sets, denoted by $\cap$, is a collection of elements that belong to both of the parent sets. In the example above:

$$A \cup B = \{1,2,3,5,7,9,11,...\} \quad \text{and} \quad A \cap B = \{1,3\}.$$

Union and intersection may also be depicted graphically in what is called a Venn diagram. A general Venn diagram for two sets is shown below.

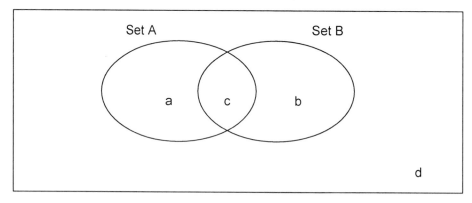

In this diagram, the quantity a represents elements that are in Set A but not Set B. Quantity b represents elements that are in Set B but not in Set A. Quantity c represents elements that are in Set A <u>and</u> Set B. Quantity d contains elements that are not in Set A and are not in Set B. When expressed in terms of union and intersection,

$$A \cup B = \{a,b,c\} \quad \text{and} \quad A \cap B = \{c\}.$$

Probability

The probability of an event is defined as $\dfrac{\text{Number of ways the event may occur}}{\text{All possible ways any event may occur}}$.

Usually probabilities are found by counting, as will be shown below. However they may also be defined in terms of geometry. Consider the Venn diagram above. The probability of an element being in Set A is $\dfrac{a+c}{a+b+c+d}$. The probability of an element being in Set B is $\dfrac{b+c}{a+b+c+d}$. The probability of an element not being in Set A and not being in Set B is $\dfrac{d}{a+b+c+d}$.

However, as stated earlier, probabilities are usually found by counting. Suppose we throw two dice at the same time and take the sum of the dots showing on the two dice. The probability of the sum of the dice equaling two is $\frac{1}{36}$. The numerator is 1 because there is only one outcome where the sum equals two (1,1). The denominator is 36 because there are 36 possible outcomes (1,1), (1,2), …, (1, 6), (2, 1), (2,2), (2,3), …. (6,6). The probability that the sum of the dice equals three is $\frac{2}{36} = \frac{1}{18}$ because there are two ways that the sum may equal three, (1,2) and (2,1).

Probability problems sometimes involve selecting certain elements from a set. For these types of problems, it is useful to know the formulas for permutations and combinations.

Permutations

We begin with an understanding of factorials. A common use of factorials is to count the number of ways that n objects can be ordered. The formula is

$$n! = n \cdot (n-1) \cdot (n-2) \ldots \cdot 3 \cdot 2 \cdot 1.$$

The exclamation point is called factorial. By definition, $0! = 1$. For example, to count the number of ways that three objects can be ordered, use three factorial:

$$3! = 3 \cdot 2 \cdot 1 = 6 \quad ABC, ACB, BAC, BCA, CAB, CBA$$

Think of this as filling three blanks: ___, ___, ___ with letters. For the first blank, there are three choices. But once the first blank is filled, there are only two remaining choices. After the first two blanks are filled there is only once choice remaining.

The general expression for permutations is the number of ways that r objects can be selected from a population of n objects, where **the order of selection matters**. The general formula for this is

$$_nP_r = \frac{n!}{(n-r)!}$$

In the example above where we were counting the number of ways three objects can be ordered, we were selecting three objects from a population of three objects, or

$$_3P_3 = \frac{3!}{(3-3)!} = \frac{3!}{0!} = 3! = 3 \cdot 2 \cdot 1 = 6$$

It is far more common to be selecting fewer objects from a larger population of objects. An example will illustrate this. Suppose there are 20 people trying out for two spots on the soccer team. The first to be selected will actually play on the team whereas the second to be selected will be an alternate. In this situation, the order of selection matters. To find the number of ways two people can be selected, use the formula:

$$_{20}P_2 = \frac{20!}{(20-2)!} = \frac{20!}{18!} = 20 \cdot 19 = 380.$$

Again, think of this as filling two blanks: ___ and ___.. There are 20 choices for the first blank, which is the person that gets to play on the team. However, once that person is chosen, there are 19 choices for the alternate spot.

Combinations

A general expression for combinations is the number of ways that r objects can be selected from a population of n objects, where **the order of selection does not matter**. The general formula for this is:

$$_nC_r = \frac{n!}{r!(n-r)!}$$

To see the difference between order mattering and order not mattering, consider once again the problem where 20 people are trying out for two spots on the soccer team. Suppose that both spots are general purpose extra players for the team. In this case, it does not matter who is chosen first and who is chosen next. The number of different outcomes for this particular problem is

$$_{20}C_2 = \frac{20!}{2!(20-2)!} = \frac{20!}{2! \cdot 18!} = \frac{20 \cdot 19}{2 \cdot 1} = 190$$

Notice that there are fewer outcomes when order does not matter. If the kids are labeled A, B, C,..., T and we are to choose two and order does not matter (combination), then the outcome AB is the same as the outcome BA. On the other hand, if the order does matter (permutation) then AB and BA are different outcomes.

Calculator tips

It is can be awkward to work with factorials, and on standardized tests speed and accuracy pays. So you might want to find these functions on your graphing calculator.

Factorials are found using the math-prb-4-! keystrokes. Practice by finding that 12! is 479,001,600.

Permutations are found by entering the value for n, then entering math-prb-2, then entering the value for r. Practice by finding that $_{10}P_2 = 30,240$.

Combinations are found by entering the value for n, then entering math-prb-3, then entering the value for r. Practice by finding that $_{10}C_3 = 120$.

Problems on sets and probability (unit 6.2)

NOTE: Use the sets: $A = \{1, 2, 3\}$ and $B = \{3, 4\}$ to answer questions 1-3.

1. The members of $A \cup B$ are:

   (A) $\{2\}$

   (B) $\{3\}$

   (C) $\{2, 3\}$

   (D) $\{1, 2, 3\}$

   (E) $\{1, 2, 3, 4\}$

2. If two numbers are chosen at random from set A, what is the probability that their sum is three?

   (A)   0

   (B)   1/2

   (C)   1/3

   (D)   2/3

   (E)   3/4

3. How many distinct ratios are formed when the numerator is taken from set A and the denominator is taken from set B?

   (A)   0

   (B)   2

   (C)   4

   (D)   6

   (E)   8

4. Sarah wants to form a rock band consisting of three guitars and one drums. She will choose them from 20 guitarists and 5 drummers. How many different rock bands could she form?

   (A)   100

   (B)   120

   (C)   1,140

   (D)   5,700

   (E)   6,840

NOTE: Use the diagram below for questions 5-6. It illustrates band and orchestra participation for the 1000 students at Central High School.

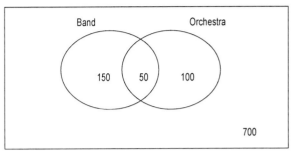

5. What is the probability that a student will play in the band?

   (A)   0.15

   (B)   0.20

   (C)   0.30

   (D)   0.50

   (E)   0.67

6. Suppose the band and orchestra each play once at different times on the same day. How many student performances will there be on that day?

   (A)   50

   (B)   200

   (C)   250

   (D)   300

   (E)   350

7. John has the difficult job of casting for the school musical. He has 10 singers on stage and will choose the best singer to play the hero, and the next best singer to play the villain. How many possible choices does John have for casting these roles in the musical?

   (A)   45

   (B)   60

   (C)   75

   (D)   90

   (E)   105

Solutions to problems on sets and probability (unit 6.2)

1. (E)

$A \cup B = \{1, 2, 3, 4\}$

2. (C)

There are $_3C_2 = 3$ ways to choose two numbers from a set of three numbers. The ways are:

$\{1,2\}$ sum=3

$\{1,3)$ sum=4

$\{2,3\}$ sum=5

The sum is three for only one way. Therefore the probability is 1/3.

3. (D)

Possible fractions are $\dfrac{1}{3}, \dfrac{1}{4}, \dfrac{2}{3}, \dfrac{2}{4}, \dfrac{3}{3}, \dfrac{3}{4}$ and they all have distinct values. So there are six of them.

4. (D)

Start with the easy one, the drummer. There are 5 choices of drummer. For guitarists, the order of selection does not matter. So

$$_{20}C_3 = \frac{20!}{3!(20-3)!} = \frac{20 \cdot 19 \cdot 18}{3 \cdot 2 \cdot 1} = 1,140 \text{ choices of}$$

guitarists. The number of possible bands is

$5(1,140) = 5,700$ bands.

5. (B)

$$\frac{150+50}{1000} = \frac{200}{1000} = \frac{1}{5} = 0.20$$

6. (E)

There will be 200 band student performances plus 150 orchestra student performances, or a total of 350 student performances.

7. (D)

$$_2P_{10} = \frac{10!}{(10-2)!} = 10 \cdot 9 = 90$$

6.3 Word problems

Almost all students have trouble with word problems, so do not feel like you are alone. The good news is that if you have been working through this book in sequence, then you have already done quite a few word problems. Earlier word problems can be found:

Type of word problem	Unit covered
Averages	2.6
Direct and inverse variation	3.6
Mixture	3.6
Percentages	2.5
Probability	6.2
Ratios	3.6
Remainders	2.3
Sequences	5.1

If you have not been moving through this book in sequence, you should backtrack through these earlier sections to make sure you can do the types of problems listed above. What follows in this section are types of word problems that are not covered earlier.

Distance, Rate, and Time

Distance rate and time problems are probably the best known of the word problems. They are based on the equation:

$$\text{Distance} = \text{Rate} \times \text{Time.}$$

Consider the following problem and its solution:

> Ellen and William started driving from the same spot at the same time, Ellen driving due East and William driving due West. If Ellen drove at 55 mph and William drove at 70 mph, how far apart were they after 5 hours have elapsed?

	Rate	x Time	=Distance
Ellen	55	5	275
William	70	5	350
TOTAL			**625**

They were 625 miles apart. The solution is found by building a table based on the equation, then filling in what is given, and calculating the rest. In this problem, it was simple to fill in the table and solve.

Tabular (equation)

Though they are popular, distance rate and time problems are not the only word problems based on a simple equation that may be used to construct a table. Consider the following:

The senior class at Central High School is selling tickets to the school play. Adult tickets cost $15 and student tickets cost $5. If the seniors raised $2,750 from ticket sales and they sold 100 adult tickets, how many student tickets did they sell?

	Price	x Quantity	= Revenue
Adult	15	100	1500
Student	5	x	
TOTAL			2750

The information given in the problem permits a table to be constructed and filled as shown above. We can deduce that $1,250 was raised from student tickets by subtracting 1500 from 2750. Now we solve a simple equation

$$5x = 1250, \quad x = 250.$$

Age

A good technique for age problems is to draw a vertical timeline and then solve. For example:

"Ellen is four times as old as William. In 20 years, Ellen will be twice as old as William. How old are they now?"

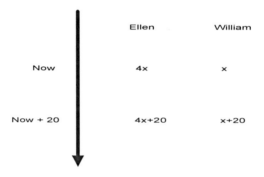

The time line enables us to find expressions for how old Ellen and William will be in 20 years. We use those expressions and the fact that in 20 years Ellen will be twice as old as William to set up an equation and solve.

$$4x + 20 = 2(x + 20), \quad 4x + 20 = 2x + 40, \quad 2x = 20, \quad x=10$$

The solution to the equation tells us that Ellen is $4x$ or 40 years old now, and William is x or 10 years old now.

Work

Work problems are best solved using substitution. Consider this problem:

"The new machine is twice as fast as the old machine. If it took 100 hours for the old machine to do a job, how long will it take to do the same job when both machines are running at the same time?"

Say that the job is to produce 1000 widgets and that the old machine took 100 hours to do the job, or 10 widgets per hour. The new machine must take 50 hours to do the job, or 20 widgets per hour. Running at the same time, the machines can process 30 widgets per hour. At that combined rate it would take $\dfrac{1000}{30} = 33\dfrac{1}{3}$ hours to do the job.

Word problems (unit 6.3)

1. Mary has to drive to her Grandmother's house and be back by 4:00. It is now 3:00 and she needs to spend 15 minutes with her Grandma. How far away is her Grandma's house if she can run this errand driving an average speed of 32 miles per hour?

(A)   12 miles

(B)   16 miles

(C)   24 miles

(D)   32 miles

(E)   36 miles

2. Alan and Bob are running a relay race where each person runs 1000 meters. If Alan's best time is 5 minutes, how fast would Bob have to run in order for their combined time to be less than 9 minutes?

(A)   100 m/min

(B)   150 m/min

(C)   200 m/min

(D)   250 m/min

(E)   300 m/min

3. One train leaves Boston and travels toward NYC at 40 mph. At the same time, a second train leaves NYC and travels toward Boston on a parallel track. After 3 hours, the trains pass each other. What is the speed of the second train if the total distance between Boston and NYC is 270 miles?

(A)   25 mph

(B)   50 mph

(C)   100 mph

(D)   125 mph

(E)   150 mph

4. You have bought 30 oz of lemonade which is 10% lemon juice. How much water must be added in order to dilute it to 3% lemon juice?

(A)   61 oz

(B)   64 oz

(C)   67 oz

(D)   70 oz

(E)   73 oz

5. Mr. Corn needs to spray his yard with 20 liters of a solution that is 30% pesticide. But the only ready-made solutions sold in his local store are 25% pesticide and 50% pesticide. How many liters of the 25% ready-made solution should he buy so that when mixed with the 50% solution, together they form 20 liters of the desired 30% solution?

(A)   8 liters

(B)   10 liters

(C)   12 liters

(D)   14 liters

(E)   16 liters

6. Dave can paint 100 blocks of wood in ten hours. Sarah is much faster, she can paint 150 blocks of wood in five hours. If they work together, how long will it take for them to paint 200 blocks of wood?

(A)   1 hour

(B)   2 hours

(C)   3 hours

(D)   4 hours

(E)   5 hours

7. Alan is 40 years older than Barbara. How much older than Barbara was Alan ten years ago?

(A)   10 years

(B)   20 years

(C)   30 years

(D)   40 years

(E)   50 years

8. Mary is three times as old as Alice. If five years from now Mary will be twice as old as Alice, how old is Alice now?

(A)   5 years old

(B)   10 years old

(C)   15 years old

(D)   20 years old

(E)   25 years old

Solutions to word problems (unit 6.3)

1. (A)

	Rate	Time	Distance
Mary	32	0.75	24

Mary would drive at 32 mph for 0.75 hours. Her roundtrip distance is 32(.75)=24 miles. So her Grandma lives 12 miles away.

2. (D)

	Rate	Time	Distance
Alan	200	5	1000
Bob	x	1000/x	1000
TOTAL			**2000**

$$5+\frac{1000}{x}<9$$

$$\frac{1000}{x}<4, \quad 1000<4x, \quad 250<x$$

3. (B)

	Rate	Time	Distance
First train	40	3	120
Second train	x	3	
TOTAL			**270**

Construct a table and fill in what is given. We can deduce that the second train has traveled 270-120=150 miles. Then solve 3x=150 or x=50 mph.

4. (D)

	Water	Juice	Total
Before	27	3	30
After	27+x	3	30+x

After adding water, the solution must be 3% juice. So

$$\frac{3}{30+x}=0.03, \quad 3=0.9+0.03x, \quad 2.1=0.03x, \quad 70=x$$

5. (E)

	Total Solution	x Strength	=Pesticide
25% solution	x	.25	.25x
50% solution	20-x	.50	10-.5x
Desired	20	.30	6

Construct a table based on the equation that the total amount of solution multiplied by its strength equals the amount of pesticide. Fill in the bottom row, for the desired 20 liters of 30% solution. Mr. Corn needs 6 liters of pesticide. Let x be the amount of 25% solution to be purchased, and complete the table.

$$.25x+(10-.5x)=6, \quad -.25x=-4, \quad x=16.$$

Mr. Corn should buy 16 liters of 25% solution and 4 liters of 50% solution.

6. (E)

Dave's rate is 100/10=10 blocks per hour. Sarah's rate is 150/5=30 blocks per hour. Their combined rate is 40 blocks per hour. At that combined rate, it will take 200/40=5 hours to paint 200 blocks of wood.

7. (D)

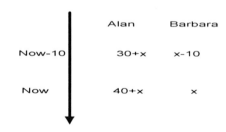

$$(30+x)-(x-10)=30+10=40$$

8. (A)

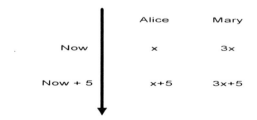

$$3x+5=2(x+5), \quad 3x+5=2x+10, \quad x=5$$

7. THE ACT® AND THE SAT® SUBJECT TESTS

The ACT® and the SAT® subject tests contain a wider range of math topics than the SAT®. Therefore, to prepare for the ACT® or either of the SAT® subject tests, more math review is needed over and above the material in the previous six chapters. The syllabus for the ACT® and the syllabus for the SAT® Level 1 subject test are very similar and, except for unit 7.10, no distinction is made between them in this chapter. Material that is needed solely for the SAT® Level 2 subject test is kept separate and is clearly indicated.

Unit 7.1 Rational numbers and scientific notation

Although neither topic has been especially popular, the standardized tests also cover rational numbers and scientific notation.

Rational Numbers

A number is rational if it can be expressed as the ratio of two integers. An easy way to recognize a rational number is to use the calculator to express it as a decimal. The number is rational if the decimal portion of the number terminates or repeats. For example:

The number	Is	Because
$2\frac{1}{2}$	Rational	(a) It can be expressed as 5/2, which is a ratio of integers; or (b) It can be expressed as 2.5, which is a terminating decimal.
0.833333333......	Rational	(a). It is a repeating decimal (the repeating group is 3); or (b) It can be expressed as 5/6.
$\sqrt{2}$	Irrational	(a) It cannot be expressed as a ratio of integer; or (b) When converted to decimal it does not repeat.
1.123123123123123......	Rational	(a) It is a repeating decimal (the repeating group is 123); or (b) It can be expressed as 1122/999.

Though the technique is not that difficult, it is not necessary to know how to take a number with a repeating decimal and convert it to a ratio of integers. Just know how to recognize a repeating group.

Scientific Notation

Scientific notation involves writing a number in the form of $a \times 10^n$, where a is a number whose absolute value is between 1 and 10, and n is an integer. That sounds pretty complicated but it is not once you've seen some examples.

Number in decimal	Number in scientific notation
$1,230,000$	1.23×10^6
0.0098723	9.8723×10^{-3}
$-12,300,000,000$	-1.23×10^{10}
-0.000009123	-9.123×10^{-6}

Problems on rational numbers and scientific notation (unit 7.1)

1. Which of the following numbers are rational?
   I.   0.101010…
   II. 6/7
   III. $\sqrt{25}$
   (A)   I only
   (B)   II only
   (C)   III only
   (D)   I and II only
   (E)   I, II, and III

2. Which of the following numbers are irrational?
   I. π
   II. $\sqrt{10}$
   III. 1.875875875….
   (A)   I only
   (B)   II only
   (C)   III only
   (D)   I and II only
   (E)   I, II, and III

3. When expressed in scientific notation 150,000 is:
   (A) 1.5×10^{-5}
   (B) 1.5×10^{5}
   (C) 0.15×10^{-5}
   (D) 15×10^{4}
   (E) 0.15×10^{6}

4. When expressed in scientific notation -0.000506 is:
   (A) 506×10^{-3}
   (B) -506×10^{3}
   (C) -5.06×10^{4}
   (D) -5.06×10^{-4}
   (E) -50.6×10^{-5}

5. What is the product of 1,350,000 and 2,300,000?
   (A) 3.105×10^{12}
   (B) 3.105×10^{-12}
   (C) 3.65×10^{5}
   (D) 3.65×10^{-5}
   (E)   27/46

6. When expressed in decimal, 6.2×10^{9} is
   (A)   0.0000000062
   (B)   0.00000000062
   (C)   6,200,000,000
   (D)   6.2000000000
   (E)   6,200,000

7. When expressed in decimal, -3.2×10^{-8} is
   (A) $-32,000,000$
   (B) -0.0000000032
   (C) -32.000000
   (D) $-.000000032$
   (E) -328

8. Find the product of $1,330,000$ multiplied by 0.22×10^{-10}.
   (A)   2,926
   (B)   2,926,000
   (C)   2,926,000,000
   (D) 2.926×10^{-3}
   (E) 2.926×10^{-5}

Solutions to problems on rational numbers and scientific notation (unit 7.1)

1. (E)

0.101010… is rational because it is repeating. The repeating group is 10.

6/7 is rational because it is the ratio of two integers.

$\sqrt{25} = 5$ is rational because 5=5/1.

2. (D)

π is irrational. When expressed in decimal it does not repeat and it does not terminate

$\sqrt{10}$ is irrational. When expressed in decimal it does not repeat and it does not terminate.

1.875875875… is rational because it is repeating. The repeating group is 875.

3. (B)

Although (D) and (E) are correct mathematically, the absolute value of the number portion must be between 1 and 10.

4. (D)

Although (E) is correct mathematically, the absolute value of the number portion must be between 1 and 10.

5. (A)

Convert each to scientific notation, then multiply.

$(1.35)(10^6)(2.3)(10^6) = 3.105 \text{ x } 10^{12}$

6. (C)

7. (D)

8. (E) .

.

Convert each number to scientific notation, then multiply

$(1.33)(10^6)(2.2)(10^{-11}) = 2.926 \text{ x } 10^{-5}$

Unit 7.2 Logarithms

It is rare when a formal definition in math is particularly useful when solving problems, but that is the case with logarithms. Take a close look at the definition:

$$\text{If } y = \log_b x \text{ then } b^y = x.$$

$$\text{If } x = b^y \text{ then } y = \log_b x$$

There are many problems that make use of this definition and you can get them easily with a little practice. But first, take a look at two special cases of the base, b.

$$\text{If } b = 10 \text{ then we write } \log x \text{ instead of } \log_{10} x$$

$$\text{If } b = e \text{ then we write } \ln x \text{ instead of } \log_e x$$

$\ln x$ is called the "natural logarithm" of x.

<u>Calculator tip:</u> The values of $\log x$ and $\ln x$ can be found on the graphing calculator, using the keys marked "LOG" and "LN". Use your calculator to find that $\log 250 = 2.4$ and $\ln 250 = 5.5$.

Practice translating the following log expressions to exponentials

$$\log x = 5 \text{ translates to } 10^5 = x, \text{ or } x = 100,000$$

$$\ln x = 3 \text{ translates to } e^3 = x, \text{ or } x = 20.08$$

$$\log_3 x = 5 \text{ translates to } 3^5 = x, \text{ or } x = 243.$$

Practice translating the following exponential expressions to logs:

$$x = 10^2 \text{ translates to } \log x = 2$$

$$x = e^3 \text{ translates to } \ln x = 3$$

$$x = 5^{12} \text{ translates to } \log_5 x = 12.$$

Laws of Logarithms

Now that you have refreshed the way logarithms work and their relationship to exponentials, it is necessary to memorize and know how to use the four laws of logarithms. They are

$$\log_b xy = \log_b x + \log_b y$$

$$\log_b \frac{x}{y} = \log_b x - \log_b y$$

$$\log_b x^y = y \log_b x$$

$$\log_b x = \frac{\log x}{\log b} = \frac{\ln x}{\ln b}$$

The last law, called the "change of base formula," is very useful when calculating the numeric value of a log that is using a base other than 10 or e.

A simple illustration of the first three laws:

$$\log_b \left(\frac{xy}{z} \right)^5 = 5 \log_b \left(\frac{xy}{z} \right) = 5 \left(\log_b xy - \log_b z \right) = 5 \left(\log_b x + \log_b y - \log_b z \right)$$

Some illustrations of the change of base formula are:

$$\log_5 170 = \frac{\log 170}{\log 5} = \frac{2.23}{0.70} = 3.2$$

or

$$\log_5 170 = \frac{\ln 170}{\ln 5} = \frac{5.14}{1.61} = 3.2$$

Problems on logarithms (unit 7.2)

1. The equation $x^3 = e^{57}$ is equivalent to:
   (A) $\log x = 19$
   (B) $\log x = 57$
   (C) $\ln x = 19$
   (D) $\ln x = 57$
   (E)   None of the above

2. The equation $5^x = y$ is equivalent to:
   I. $\log_5 x = y$

   II. $\log_5 y = x$

   III. $\dfrac{\log y}{\log 5} = x$

   (A)   I only
   (B)   II only
   (C)   III only
   (D)   I and III only
   (E)   II and III only

3. If $x = \left(\dfrac{a^2 b^3}{a + c^2}\right)^5$, then $\log x$ is:

   (A) $5\log a + 15\log b - 10\log c$

   (B) $10\log a + 15\log b - 5\log(a + c^2)$

   (C) $5\log a + 15\log b - 5\log c$

   (D) $15\log a + 15\log b + 10\log c$

   (E) None of the above

4. If $x = \log_3 45 - \log_3 5$, then x is equal to
   (A)   -1
   (B)   0
   (C)   1
   (D)   2
   (E)   3

5. If $x = 15\log_{15} 3 + 15\log_{15} 5$ then x is equal to:
   (A)   0
   (B)   1
   (C)   3
   (D)   5
   (E)   15

6. If $5^x = 125^{x+4}$ then x is equal to:
   (A)   -6
   (B)   -2
   (C)   2
   (D)   6
   (E)   10

7. If $\log_3(2x - 1) = 2$ then x is equal to:
   (A)   3
   (B)   3.5
   (C)   4
   (D)   5
   (E)   6

8. If $\log x + \log 2 = 2$ then x is equal to:
   (A)   5
   (B)   10
   (C)   25
   (D)   50
   (E)   100

9. Without a calculator, find the value of $\dfrac{\ln 8}{\ln 2}$
   (A)   2
   (B)   3
   (C)   4
   (D)   6
   (E)   8

Solutions to problems on logarithms (unit 7.2)

1. (C)

$x^3 = e^{57}$

$\ln x^3 = 57, \ 3\ln x = 57, \ \ln x = \dfrac{57}{3} = 19$

2. (E)

By definition, I is false and II is true. Use the change of base formula to see that if II is true then III is true.

$x = \log_5 y = \dfrac{\log y}{\log 5}$

3. (B)

$x = \left(\dfrac{a^2 b^3}{a + c^2}\right)^5, \ \log x = 5\log\left(\dfrac{a^2 b^3}{a + c^2}\right)$

$\log x = 5\left(\log a^2 b^3 - \log\left(a + c^2\right)\right)$

$\log x = 5\left(2\log a + 3\log b - \log\left(a + c^2\right)\right)$

$\log x = 10 \log a + 15 \log b - 5\log(a + c^2)$

4. (D)

$x = \log_3 45 - \log_3 5 = \log_3\left(\dfrac{45}{5}\right) = \log_3 9$

$x = \log_3 9$ translates to $3^x = 9, \ x = 2$

5. (E)

$x = 15\log_{15} 3 + 15\log_{15} 5$

$x = 15(\log_{15} 3 + \log_{15} 5)$

$x = 15\log_{15}(3 \cdot 5) = 15\log_{15} 15 = 15 \cdot 1 = 15$

6. (A)

$5^x = 125^{x+4}$

$5^x = \left(5^3\right)^{x+4} = 5^{3x+12}$

$x = 3x + 12, \ x = -6$

7. (D)

$\log_3\left(2x - 1\right) = 2$

$3^2 = 2x - 1$

$9 = 2x - 1, \ x = 5$

8. (D)

$\log x + \log 2 = 2$

$\log 2x = 2$

$10^2 = 2x, \ 100 = 2x, \ x = 50$

9. (B)

Use the change of base formula

$\dfrac{\ln 8}{\ln 2} = \log_2 8$

$x = \log_2 8, \ 2^x = 8, \ x = 3$

Unit 7.3 Matrix algebra

A matrix is a collection of numbers arranged into rows and columns. A matrix is denoted by a capital letter, and its members are denoted by lower case letters. There are only a few basic operations that you need to know.

In the material below, we will use these two general expressions for the matrices A and B

$$A = \begin{pmatrix} a_{1,1} & a_{1,2} \\ a_{2,1} & a_{2,2} \end{pmatrix} \quad B = \begin{pmatrix} b_{1,1} & b_{1,2} \\ b_{2,1} & b_{2,2} \end{pmatrix}$$

It will also help to use a specific example, so we will let

$$A = \begin{pmatrix} 1 & 2 \\ 3 & 4 \end{pmatrix} \quad B = \begin{pmatrix} 5 & 6 \\ 7 & 8 \end{pmatrix}$$

Multiplication by a scalar

If k is a scalar, then $kA = \begin{pmatrix} ka_{1,1} & ka_{1,2} \\ ka_{2,1} & ka_{2,2} \end{pmatrix} \quad kB = \begin{pmatrix} kb_{1,1} & kb_{1,2} \\ kb_{2,1} & kb_{2,2} \end{pmatrix}$

Suppose k=5. Then $kA = \begin{pmatrix} 5 \cdot 1 & 5 \cdot 2 \\ 5 \cdot 3 & 5 \cdot 4 \end{pmatrix} = \begin{pmatrix} 5 & 10 \\ 15 & 20 \end{pmatrix} \quad kB = \begin{pmatrix} 5 \cdot 5 & 5 \cdot 6 \\ 5 \cdot 7 & 5 \cdot 8 \end{pmatrix} = \begin{pmatrix} 25 & 30 \\ 35 & 40 \end{pmatrix}$

Addition and subtraction

Addition and subtraction can only take place when both matrices have the same dimensions. When this occurs, simply add or subtract corresponding elements:

$$A \pm B = \begin{pmatrix} a_{1,1} & a_{1,2} \\ a_{2,1} & a_{2,2} \end{pmatrix} \pm \begin{pmatrix} b_{1,1} & b_{1,2} \\ b_{2,1} & b_{2,2} \end{pmatrix} = \begin{pmatrix} a_{1,1} \pm b_{1,1} & a_{1,2} \pm b_{1,2} \\ a_{2,1} \pm b_{2,1} & a_{2,2} \pm b_{2,2} \end{pmatrix}$$

Suppose we want to calculate **5A - 2B**. This would be

$$5A - 2B = \begin{pmatrix} 5 \cdot 1 & 5 \cdot 2 \\ 5 \cdot 3 & 5 \cdot 4 \end{pmatrix} - \begin{pmatrix} 2 \cdot 5 & 2 \cdot 6 \\ 2 \cdot 7 & 2 \cdot 8 \end{pmatrix} = \begin{pmatrix} 5 & 10 \\ 15 & 20 \end{pmatrix} - \begin{pmatrix} 10 & 12 \\ 14 & 16 \end{pmatrix} = \begin{pmatrix} 5-10 & 10-12 \\ 15-14 & 20-16 \end{pmatrix} = \begin{pmatrix} -5 & -2 \\ 1 & 4 \end{pmatrix}$$

Multiplication

Multiplication is the most difficult of the matrix operations. Remember that multiplication may only take place when the inner dimensions of the two matrices are identical. The dimensions of the resulting product will have the outer dimensions of the matrices that have been multiplied. This sounds complicated, but it's not.

Suppose **C** is a matrix with dimensions (2x3) and **D** is a matrix with dimensions (3x2). The product **CD** will have dimensions (2x2) because:

$$CD = \underset{(2x3)}{\mathbf{C}} \cdot \underset{(3x2)}{\mathbf{D}} = \underset{(2x2)}{\mathbf{E}}$$

The inner dimensions sort of "cancel out" leaving the outer dimensions for the result.

The product DC will have dimensions (3x3) because:

$$DC = \underset{(3x2)}{\mathbf{D}} \cdot \underset{(2x3)}{\mathbf{C}} = \underset{(3x3)}{\mathbf{F}}$$

It's important to understand dimensions before actually doing some problems because dimensions are used to guide the way things are set up. Notice that **E** and **F** are different matrices with different dimensions, proving that in the world of matrices $\mathbf{C} \cdot \mathbf{D} \neq \mathbf{D} \cdot \mathbf{C}$. So now we are ready to take a look at how multiplication actually works. Suppose:

$$\mathbf{C} = \begin{pmatrix} 1 & 3 & -1 \\ 0 & -2 & 1 \end{pmatrix} \quad \mathbf{D} = \begin{pmatrix} 1 & 2 \\ 0 & -1 \\ -2 & 1 \end{pmatrix}$$

Then

$$\mathbf{CD} = \begin{pmatrix} 1\cdot1+3\cdot0+-1\cdot-2 & 1\cdot2+3\cdot-1+-1\cdot1 \\ 0\cdot1+-2\cdot0+1\cdot-2 & 0\cdot2+-2\cdot-1+1\cdot1 \end{pmatrix} = \begin{pmatrix} 1+0+2 & 2-3-1 \\ 0+0-2 & 0+2+1 \end{pmatrix} = \begin{pmatrix} 3 & -2 \\ -2 & 3 \end{pmatrix}$$

Notice that the first row and first column of the resulting product matrix consists of the first row of **C** multiplied by the first column of **D**. The first row and second column consists of the first row of **C** multiplied by the second column of **D**, etc.

Now take a look at **DC**. Notice that $\mathbf{DC} \neq \mathbf{CD}$.

$$\mathbf{DC} = \begin{pmatrix} 1\cdot1+2\cdot0 & 1\cdot3+2\cdot-2 & 1\cdot-1+2\cdot1 \\ 0\cdot1-1\cdot0 & 0\cdot3-1\cdot-2 & 0\cdot-1-1\cdot1 \\ -2\cdot1+1\cdot0 & -2\cdot3+1\cdot-2 & -2\cdot-1+1\cdot1 \end{pmatrix} = \begin{pmatrix} 1+0 & 3-4 & -1+2 \\ 0-0 & 0+2 & 0-1 \\ -2+0 & -6-2 & 2+1 \end{pmatrix} = \begin{pmatrix} 1 & -1 & 1 \\ 0 & 2 & -1 \\ -2 & -8 & 3 \end{pmatrix}$$

Problems on matrix algebra (unit 7.3)

Use these matrices to solve problems 1-5 below:

$$A = \begin{pmatrix} 1 & 0 \\ 2 & -1 \end{pmatrix} \quad B = \begin{pmatrix} 0 & 2 \\ 1 & 1 \end{pmatrix} \quad C = \begin{pmatrix} 1 & 2 & -1 \\ 2 & 0 & 1 \end{pmatrix}$$

1. The sum of **A** and **B** is:

(A) $\begin{pmatrix} 1 & -2 \\ 1 & -2 \end{pmatrix}$

(B) $\begin{pmatrix} -1 & 2 \\ -1 & -2 \end{pmatrix}$

(C) $\begin{pmatrix} 1 & 2 \\ 3 & 0 \end{pmatrix}$

(D) $\begin{pmatrix} 3 & 0 \\ 3 & 0 \end{pmatrix}$

(E)   Cannot be determined

2. The value of **B − 2A** is:

(A) $\begin{pmatrix} 0 & 2 \\ -3 & 0 \end{pmatrix}$

(B) $\begin{pmatrix} -2 & 2 \\ -3 & 3 \end{pmatrix}$

(C) $\begin{pmatrix} -1 & 2 \\ -1 & -2 \end{pmatrix}$

(D) $\begin{pmatrix} 1 & 2 \\ 3 & 0 \end{pmatrix}$

(E)   Cannot be determined

3. The value of **BA** is:

(A) $\begin{pmatrix} 4 & -2 \\ 3 & -1 \end{pmatrix}$

(B) $\begin{pmatrix} 0 & 2 \\ 1 & 3 \end{pmatrix}$

(C) $\begin{pmatrix} 4 & 2 \\ 3 & -1 \end{pmatrix}$

(D) $\begin{pmatrix} 0 & 2 \\ 3 & 3 \end{pmatrix}$

(E)   Cannot be determined

4. The value of **A + C** is:

(A) $\begin{pmatrix} 2 & 2 & -1 \\ 4 & -1 & 1 \end{pmatrix}$

(B) $\begin{pmatrix} 1 & 4 & -1 \\ 3 & 1 & 1 \end{pmatrix}$

(C) $\begin{pmatrix} 1 & 3 & -1 \\ 2 & 2 & 0 \end{pmatrix}$

(D) $\begin{pmatrix} 1 & 2 & 1 \\ 2 & 1 & 2 \end{pmatrix}$

(E)   Cannot be determined

5. The value of **AC** is:

(A) $\begin{pmatrix} 4 & 0 & 2 \\ 3 & 2 & 0 \end{pmatrix}$

(B) $\begin{pmatrix} 4 & 0 & 1 \\ 3 & 2 & 0 \end{pmatrix}$

(C) $\begin{pmatrix} 1 & 2 & 1 \\ 0 & 3 & -3 \end{pmatrix}$

(D) $\begin{pmatrix} 1 & 2 & -1 \\ 0 & 4 & -3 \end{pmatrix}$

(E)   Cannot be determined

6. If $\begin{pmatrix} 1 & 0 \\ 2 & -1 \end{pmatrix}\begin{pmatrix} x \\ y \end{pmatrix} = \begin{pmatrix} 3 \\ -6 \end{pmatrix}$, find the value of y.

(A)   -12

(B)   -6

(C)   0

(D)   6

(E)   12

Solutions to problems on matrix algebra (unit 7.3)

1. (C)

$$\mathbf{A} + \mathbf{B} = \begin{pmatrix} 1+0 & 0+2 \\ 2+1 & -1+1 \end{pmatrix} = \begin{pmatrix} 1 & 2 \\ 3 & 0 \end{pmatrix}$$

2. (B)

$$\mathbf{B} - 2\mathbf{A} = \begin{pmatrix} 0 & 2 \\ 1 & 1 \end{pmatrix} - \begin{pmatrix} 2 & 0 \\ 4 & -2 \end{pmatrix} = \begin{pmatrix} -2 & 2 \\ -3 & 3 \end{pmatrix}$$

3. (A)

$$\mathbf{BA} = \begin{pmatrix} 0 \cdot 1 + 2 \cdot 2 & 0 \cdot 0 + 2 \cdot -1 \\ 1 \cdot 1 + 1 \cdot 2 & 1 \cdot 0 + 1 \cdot -1 \end{pmatrix} = \begin{pmatrix} 4 & -2 \\ 3 & -1 \end{pmatrix}$$

4. (E)

These matrices cannot be added because they have different dimensions.

5. (D)

$$\mathbf{AC} = \begin{pmatrix} 1 \cdot 1 + 0 \cdot 2 & 1 \cdot 2 + 0 \cdot 0 & 1 \cdot -1 + 0 \cdot 1 \\ 2 \cdot 1 - 1 \cdot 2 & 2 \cdot 2 - 1 \cdot 0 & 2 \cdot -1 - 1 \cdot 1 \end{pmatrix} = \begin{pmatrix} 1 & 2 & -1 \\ 0 & 4 & -3 \end{pmatrix}$$

6. (E)

$$\begin{pmatrix} 1 & 0 \\ 2 & -1 \end{pmatrix} \begin{pmatrix} x \\ y \end{pmatrix} = \begin{pmatrix} 3 \\ -6 \end{pmatrix}$$

$x = 3$

$2x - y = -6$

$2(3) - y = -6, \ y = 12$

Unit 7.4 Quadratic equations

For the SAT®, you need to know how to factor quadratic equations and solve for the zeros (roots) of the equation, and you need to have some knowledge of the parabolas that are obtained when the quadratic equation is graphed (see unit 5.4) But for the ACT® and SAT® subject tests, there is more that you are expected to know about quadratic equations.

Calculator Tip: When solving problems involving quadratic equations, do not forget that you can use your graphing calculator. The calculator can graph the parabola, find the zeros of the equation (use *2nd-calc-zero*), and find its vertex (if the parabola opens up, use *2nd-calc-minimum* if the parabola opens down, use *2nd-calc-maximum*). The material below demonstrates how to accomplish these tasks by hand.

Equations in Standard Form

A quadratic equation in standard form is $y = ax^2 + bx + c$.

The graph of a quadratic equation is a parabola. The parabola opens up if $a > 0$ and the parabola opens down if $a < 0$. The parabola is symmetric about its axis of symmetry, which is an invisible line running vertically through the vertex. The equation of the axis of symmetry is $x = \dfrac{-b}{2a}$. This is illustrated in the figures below.

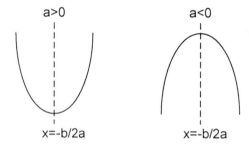

The zeros or roots of the equation are obtained by setting it equal to zero, i.e., the roots are the values of x for which $ax^2 + bx + c = 0$. The roots are where the graph of the parabola crosses the x-axis.

There are several methods to find the roots. One method is to factor (see unit 3.5) and set each factor to zero. Another method is to use the quadratic formula, which you are expected to memorize:

$$x = \frac{-b \pm \sqrt{b^2 - 4ac}}{2a}.$$

You are also expected to memorize that the sum of the two roots is $\dfrac{-b}{a}$ and that the product of the roots is $\dfrac{c}{a}$.

By graphing the equation, you can see what kinds of roots exist. If the parabola never crosses the x-axis then the equation has two imaginary roots. If the parabola crosses the x-axis at two points, then the equation has two real roots. If the vertex of the parabola lies on the x-axis then the equation has one real root (also called a double root).

Instead of graphing, the kinds of roots can be determined by the sign of the discriminant, $b^2 - 4ac$. When the discriminant is positive, there are two real roots. When the discriminant is negative, there are two imaginary roots. When the discriminant is zero there is one real root. You are expected to be able to determine the number and type of roots by using the discriminant, rather than solving the equation for its roots.

Equations in Vertex Form

If the goal is to graph the parabola, it is easier to do so when the equation is written in vertex form rather than standard form. The vertex form can look like

$$y\text{-}k = a(x\text{-}h)^2 \text{ or } y = a(x\text{-}h)^2 + k.$$

In either case, the vertex is (h,k) and the axis of symmetry is the equation $x = h$. This is illustrated in the figures below.

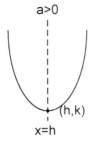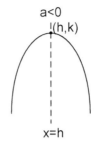

You should be prepared to see quadratic equations in either standard form or vertex form. Remember how to convert an equation from standard form to vertex form using the technique called "completing the square?" This technique is illustrated below.

To find the vertex of the equation $y = x^2 - 12x + 46$:

$y = x^2 - 12x + 46$	the original equation
$y - 46 = x^2 - 12x$	isolate the x-terms
$y - 46 + 36 = x^2 - 12x + 36$	add the constant $\left(\dfrac{b}{2}\right)^2 = \left(\dfrac{-12}{2}\right)^2 = (-6)^2 = 36$
$y + 10 = (x - 6)^2$	simplify
$(6, \text{-}10)$	coordinates of the vertex

Problems on quadratic equations (unit 7.4)

1. For the quadratic equation $ax^2 + bx + c$, when the sum of the two roots is added to the product of the two roots, the result is:

(A) $\dfrac{-b}{2a}$

(B) $\dfrac{c}{a}$

(C) $\dfrac{c-b}{a}$

(D) $\dfrac{b-c}{a}$

(E) $\dfrac{-c}{a}$

2. Which of the points below lies on the axis of symmetry for the parabola with the equation of $y = 3x^2 + 12x - 15$?

(A) (-4,3)

(B) (-2,2)

(C) (0,6)

(D) (2,5)

(E) (4,4)

3. For which of the equations below is the value of the discriminant equal to -23?

(A) $y = x^2$

(B) $y = x^2 - x - 3$

(C) $y = -3x^2 - 2x + 7$

(D) $y = x^2 - 3x + 2.25$

(E) $y = 3x^2 - 5x + 4$

4. A parabola never crosses the x-axis if its discriminant is:

(A) positive

(B) negative

(C) zero

(D) zero or positive

(E) zero or negative

5. One of the solutions to the equation $x^2 - 3x + 2.25 = 0$ is:

(A) 0.5

(B) 1.0

(C) 1.5

(D) 2.0

(E) 2.5

6. When the equation $y = x^2 - 6x - 3$ is written in vertex form, it becomes:

(A) $y = (x+3)^2 - 12$

(B) $y = (x+3)^2 + 12$

(C) $y = (x-3)^2 - 12$

(D) $y = (x-3)^2 + 12$

(E) $y = (x-3)^2 - 6$

7. If a parabola intersects the x-axis at the points (-3,0) and (3,0), then the equation of the parabola could be:

I. $y = x^2 - 9$

II. $y = -x^2 + 9$

III. $y = (x+3)^2$

(A) I only

(B) II only

(C) III only

(D) I and II only

(E) I and III only

Solutions to problems on quadratic equations (unit 7.4)

1. (C)

The sum of the roots is –b/a, whereas the product of the roots is c/a. When the sum of the roots is added to the product of the roots we get

$$\frac{-b}{a} + \frac{c}{a} = \frac{c-b}{a}$$

2. (B)

In the equation $y = 3x^2 + 12x - 15$, a=3 and b=12. So the axis of symmetry is –b/2a=-12/6 = -2. Any point with an x-coordinate of -2 could lie on the axis of symmetry.

3. (E)

Calculate the discriminant for each equation.

      A. D=0-0=0

      B. D=1-4(-3)=13

      C. D=4-4(-21)=88

      D. D=9-4(2.25)=0

      E. D=25-4(12)=-23

4. (B)

If the discriminant is negative, there are no real roots and the parabola never crosses the x-axis.

5. (C)

Use the quadratic formula. For $x^2 - 3x + 2.25 = 0$

$$x = \frac{3 \pm \sqrt{9 - 4(2.25)}}{2} = \frac{3 \pm \sqrt{9-9}}{2} = \frac{3}{2}$$

6. (C)

Complete the square.

$$y = x^2 - 6x - 3$$
$$y + 3 = x^2 - 6x$$
$$y + 3 + 9 = x^2 - 6x + 9$$
$$y + 12 = (x-3)^2$$
$$y = (x-3)^2 - 12$$

7. (D)

The first two equations cross the x-axis at the points indicated, whereas the third equation crosses the

x-axis at only one point (-3,0).

Unit 7.5 Inverse functions

This unit builds on Unit 5.6, which covered the definition of functions and compound functions. If you are unsure about these topics, please review Unit 5.6 before proceeding.

Recall that in order for a relation to be a function, it must pass the vertical line test -- each value in the domain can map to one and only one value in the range. Thus $f(x) = x^2$ is a function. However, $f(x) = \pm\sqrt{x}$ is not a function because each value of x maps to two y-values, e.g., 9 maps to 3 and -3.

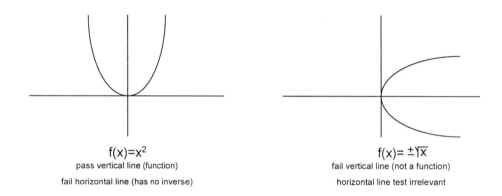

f(x)=x²	f(x)= ±√x̄
pass vertical line (function)	fail vertical line (not a function)
fail horizontal line (has no inverse)	horizontal line test irrelevant

In order for a function to have an inverse, it must pass the horizontal line test, where each value in the range maps to one and only one value in the domain. For example, the relation $f(x) = x^2$ is a valid function (it passes the vertical line test) but it has no inverse (it fails the horizontal line test).

Finding the inverse $f^{-1}(x)$ of a function $f(x)$

To find the inverse, called $f^{-1}(x)$, of a given function, called $f(x)$, first make sure that $f(x)$ is a function (passes the vertical line test) and that it has an inverse (passes the horizontal line test). Then follow these steps:

- Replace f(x) with y.
- Switch x and y
- Solve for y

For example,

$$f(x) = \frac{x}{5} + 3 \qquad\qquad \text{given the function f(x)}$$

$$y = \frac{x}{5} + 3 \qquad\qquad \text{replace f(x) with y}$$

$$x = \frac{y}{5} + 3 \qquad\qquad \text{switch x and y}$$

$$y = 5x - 15 \qquad\qquad \text{solve for y}$$

$$f^{-1}(x) = 5x - 15$$

Proving that a given function is an inverse function

To prove that a given function $f^{-1}(x)$ is the inverse of another function $f(x)$, simply examine the compound functions to determine whether

$$f\left(f^{-1}(x)\right) = x \text{ and } f^{-1}\left(f(x)\right) = x.$$

Using the example from above, we wish to determine whether $f(x) = \dfrac{x}{5} + 3$ and $f^{-1}(x) = 5x - 15$ really are inverses of each other.

$$f\left(f^{-1}(x)\right) = f(5x - 15) = \frac{5x - 15}{5} + 3 = x - 3 + 3 = x$$

$$f^{-1}\left(f(x)\right) = f^{-1}\left(\frac{x}{5} + 3\right) = 5\left(\frac{x}{5} + 3\right) - 15 = x + 15 - 15 = x$$

So now it is possible to conclude that the function $5x - 15$ really is the inverse of the function $\dfrac{x}{5} + 3$.

Problems on inverse functions (unit 7.5)

1. The inverse of $f(x) = \dfrac{2x}{3} + 8$ is:

    (A) $1.5x - 4$

    (B) $1.5x - 8$

    (C) $1.5x - 12$

    (D) $1.5x - 16$

    (E)   An inverse does not exist

2. The inverse of $f(x) = x^2 + 6x + 9$ is:

    (A) $y = x^2$

    (B) $y = x^2 - x - 3$

    (C) $y = -3x^2 - 2x + 7$

    (D) $y = x^2 - 3x + 2.25$

    (E)   An inverse does not exist

3. The inverse of $f(x) = \dfrac{x^3 + 2}{5}$ is:

    (A) $\sqrt[3]{2x + 5}$

    (B) $\sqrt[3]{5x + 2}$

    (C) $\sqrt[3]{\dfrac{x - 5}{2}}$

    (D) $\sqrt[3]{5x - 2}$

    (E)   An inverse does not exist

4. The inverse of $f(x) = |x - 5| + 3$ is:

    (A) $x + 2$

    (B) $x - 2$

    (C) $x - 8$

    (D) $8 - x$

    (E)   An inverse does not exist

5. If the inverse of a function is $\dfrac{x + 2}{2}$ then the original function is:

    (A) $\dfrac{x + 2}{2}$

    (B) $\dfrac{x - 2}{2}$

    (C) $x - 1$

    (D) $2x - 2$

    (E)   Cannot be determined

Answers to problems on inverse functions (unit 7.5)

1. (C)

$$f(x) = \frac{2x}{3} + 8$$

$$x = \frac{2y}{3} + 8$$

$$y = f^{-1}(x) = \frac{3(x-8)}{2} = 1.5x - 12$$

2. (E)

$y = x^2 + 6x + 9$ fails the horizontal line test and therefore has no inverse.

3. (D)

$$f(x) = \frac{x^3 + 2}{5}$$

$$x = \frac{y^3 + 2}{5}$$

$$y = \sqrt[3]{5x - 2}$$

4. (E)

$y = |x - 5| + 3$ fails the horizontal line test and therefore has no inverse.

5. (D)

To find the original function, take the inverse of the inverse. Or you can solve this by taking the inverse of each of the answers to see which one matches the inverse that was given.

$$f^{-1}(x) = \frac{x+2}{2}$$

$$x = \frac{y+2}{2}$$

$$f(x) = y = 2x - 2$$

Unit 7.6   More on sequences

Yes, there is more you need to know about sequences. Please review unit 5.1 on sequences before doing this unit.

Recall that a sequence is a collection of terms, denoted by a_1, a_2, ..., a_n and a series is the sum of consecutive terms of a sequence. There are two types of sequences, arithmetic and geometric.

Arithmetic sequences

Arithmetic sequences have the common difference d, i.e., the difference between adjacent terms is always the same. The expression for the nth term of an arithmetic sequence is

$$a_n = a_1 + (n-1)d$$

The sum of n consecutive terms of an arithmetic sequence is

$$s_n = \frac{n(a_1 + a_n)}{2},$$

where a_1 is the first term to be summed, a_n is the last term to be summed, and n is the number of terms to be summed.. For example, consider the sequence: 3,11,19, ….. The sum of the first 100 terms of the sequence would be:

$$s_{100} = \frac{100(a_1 + a_{100})}{2}$$

From $a_{100} = 3 + (100-1)8 = 795$ we get

$$s_{100} = \frac{100(a_1 + a_{100})}{2} = \frac{100(3+795)}{2} = 39,900$$

Geometric sequences

Geometric sequences have the common ratio r, i.e., the ratio of any term to the previous term is always the same. The expression for the nth term of a geometric sequence is

$$a_n = a_1 r^{n-1}$$

The sum of n consecutive terms of a geometric sequence is

$$s_n = \frac{a_1(1-r^n)}{1-r},$$

where a_1 is the first term to be summed and n is the number of terms to be summed. For example, consider the sequence 3,27,243, etc. The sum of the first 5 terms would be:

$$s_5 = \frac{3(1-9^5)}{1-9} = \frac{3(-59,048)}{-8} = 22,143$$

The sums above are called partial sums or finite sums because they involve adding only n consecutive terms in the sequence. For infinite geometric sequences, if $|r| < 1$ then the sum of all the elements of the sequence will converge as n becomes larger and larger. Specifically, the sum will converge to:

$$S_\infty = \frac{a_1}{1-r}, \text{ when } |r| < 1$$

A special case of geometric sequences is especially useful to know, not only for standardized tests but also it is very useful in the real world. The special case is money compounding over time. If A dollars is invested (say in a savings account) and interest is allowed to compound annually at the rate of r per year, after t years the investment will be worth $A(1+r)^t$. In the real world, interest on an investment usually compounds more often than once per year. If the investment compounds n times per year at an annual interest rate of r per year, after t years the

investment will be worth $A\left(1+\dfrac{r}{n}\right)^{nt}$.

A simple example will illustrate the power of compound interest. At an annual rate of 5% per year, after 20 years an investment of \$25,000 would be worth $25,000(1+0.05)^{20} = 66,332$ when compounded annually. If that same

\$25,000 investment were to compound monthly, after 20 years it would be worth $25,000\left(1+\dfrac{.05}{12}\right)^{240} = 67,816$.

Problems on sequences (unit 7.6)

1. For the sequence -3,-9,-15, … what is the value of the 50ᵗʰ term?

    (A)   -291

    (B)   -294

    (C)   -297

    (D)   -300

    (E)   -303

2. For the sequence   -3,-9,-15, …what is the sum of the first 50 terms?

    (A)   -7,425

    (B)   -7,500

    (C)   -14,850

    (D)   -15,000

    (E)   -15,150

3. For the sequence 5,- 7.5, 11.25, -16.875, … what is the 15ᵗʰ term?

    (A)   -2189

    (B)   -1460

    (C)   292

    (D)   1460

    (E)   2189

4. For the sequence 5,- 7.5, 11.25, -16.875, … what is the sum of the first 15 terms?

    (A)   582

    (B)   873

    (C)   878

    (D)   4379

    (E)   8780

5. What is the value of the infinite sum 9+3+1+… ?

    (A)   13.5

    (B)   27.0

    (C)   40.5

    (D)   54.0

    (E)   Does not converge

6. What is the sum of the odd integers between 100 and 300?

    (A)   4,900

    (B)   9,800

    (C)   10,000

    (D)   19,800

    (E)   20,000

7. John has an ant farm whose population doubles every 5 years. If he started the ant farm 20 years ago and there are now 160,000 ants, how many ants did he have when he started?

    (A)   1

    (B)   2,500

    (C)   5,000

    (D)   7,500

    (E)   10,000

8. Which of the models below fits the following sequence: 4, 7, 12, 19, 28, …?

    (A) $a_n = n^3 - 1$

    (B) $a_n = 2^{n+1}$

    (C) $a_n = 4 + (n-1)3$

    (D) $a_n = n^2 + 3$

    (E) $a_n = 3n + 1$

9. Amy wants to invest $1,000 for a period of 10 years, but she cannot decide where to place her investment. Bank A will pay 5.5% interest compounded annually. Bank B will pay 5% interest compounded quarterly. Which is the better deal?

    (A)   Bank A is better

    (B)   Bank B is better

    (C)   Both deals are the same

    (D)   Cannot be determined

Solutions to problems on sequences (unit 7.6)

1. (C)

This is arithmetic, with d=-6.

$$a_{50} = -3 + (50-1)(-6) = -297$$

2. (B)

$$s_{50} = \frac{50(-3-297)}{2} = -7500$$

3. (D)

This is geometric, with r=-1.5.

$$a_{15} = 5(-1.5)^{15-1} = 1459.646$$

4. (C)

This is geometric, with r=-1.5.

$$s_{15} = \frac{5\left(1-(-1.5)^{15}\right)}{1--1.5} = \frac{5(1+437.89)}{2.5} = 877.79$$

5. (A)

This is geometric, with r=1/3.

$$s_{\infty} = \frac{9}{1-\frac{1}{3}} = 13.5$$

6. (E)

This is an arithmetic sequence: 101, 103, …, 299 with d=2 and n=100.

$$s_{100} = \frac{100(101+299)}{2} = 20,000$$

7. (E)

This is a geometric sequence, with r=2. Twenty years later would be the fifth element in the sequence:

$$a_5 = a_1(2)^{5-1}, \quad 160{,}000 = 16a_1, \quad a_1 = 10,000$$

8. (D)

Sequences can sometimes be neither arithmetic nor geometric. $a_n = n^2 + 3$ is the only model that fits all of the data given.

9. (A)

5.5% compounded annually for 10 years:

$$1000(1.055)^{10} = 1708.14$$

5.0% compounded quarterly for 10 years:

$$1000\left(1+\frac{.05}{4}\right)^{10\cdot4} = 1000(1.0125)^{40} = 1643.62$$

The better deal is the first one, Bank A.

Unit 7.7 Complex numbers

Suppose we want to solve the equation

$$x^2 + 9 = 0 .$$

We would perform two steps and then be stuck:

$$x^2 = \text{-}9$$
$$x = \pm\sqrt{\text{-}9}$$

We are stuck because a negative number cannot have a real square root. So to get around this, mathematicians invented i which is called the imaginary number. By definition, $i = \sqrt{\text{-}1}$ and $i^2 = \text{-}1$. So now we can finish the problem by writing:

$$x = \pm\sqrt{\text{-}9} = \pm\sqrt{9} \cdot \sqrt{\text{-}1} = \pm 3i$$

The expression $3i$ is called an imaginary number. When an imaginary number is combined with an integer, it is called a complex number. For example $5 - 3i$ is a complex number. Complex numbers take the general form of $a + bi$, where a and b are integers, though not necessarily positive integers. When $a = 0$ we have an imaginary number, and when $b = 0$ we have an integer.

Operations on complex numbers

Now that you understand what complex numbers are and one of the reasons why they were invented, you need to be able to perform three operations: addition, subtraction, and multiplication.

Suppose we have two complex numbers, $a + bi$ and $c + di$. Then:

$$(a+bi)+(c+di)=(a+c)+(b+d)i$$
$$(a+bi)-(c+di)=(a-c)+(b-d)i$$

For example

$$(2+i)+(3-2i)=2+3+i-2i=5-i$$
$$(2+i)-(3-2i)=2-3+i+2i=\text{-}1+3i$$

As you can see above, addition and subtraction are what you would expect. Treat i as though it were a variable. The same applies to multiplication—the complex numbers are FOILed:

$$(a+bi)(c+di)=ac+adi+bci+bdi^2=(ac-bd)+(ad+bc)i$$

For example:

$$(2+i)(3-2i)=2\cdot 3-4i+3i-2i^2=6-i-2(\text{-}1)=8-i$$

Division involving complex numbers (the use of conjugates)

You must know how to simplify expressions with complex numbers in the denominator. This is done using conjugates. The conjugate of $a + bi$ is $a - bi$. In other words, to find the conjugate of a complex number just change the sign of the imaginary portion of the number.

For example, the conjugate of $3 - 5i$ is $3 + 5i$. We use this fact to simplify fractions, as in the following

$$\frac{2}{3-5i} = \left(\frac{2}{3-5i}\right)\left(\frac{3+5i}{3+5i}\right) = \frac{2(3+5i)}{9-15i+15i-25i^2} = \frac{2(3+5i)}{9+25} = \frac{2(3+5i)}{34} = \frac{3+5i}{17}$$

Powers of i

A favorite type of problem involves powers of i. These follow a pattern that is easy to find, once you get the hang of it.

$$i = i$$
$$i^2 = -1$$
$$i^3 = i \cdot i^2 = -i$$
$$i^4 = \left(i^2\right)^2 = (-1)^2 = 1$$

Powers of i will always equal 1 or -1 when the power is even; and i or $-i$ when the power is odd. Some examples:

$$i^{18} = \left(i^2\right)^9 = (-1)^9 = -1$$
$$i^{52} = \left(i^2\right)^{26} = (-1)^{26} = 1$$
$$i^{17} = i^{16} \cdot i = \left(i^2\right)^8 \cdot i = (-1)^8 \cdot i = i$$
$$i^{51} = i^{50} \cdot i = \left(i^2\right)^{25} \cdot i = (-1)^{25} \cdot i = -i$$

Problems on complex numbers (unit 7.7)

1. The sum of $2-3i$ and its conjugate is:

    (A)   4

    (B) $4-6i$

    (C) $4+6i$

    (D) $-6i$

    (E) $6i$

2. The product of $2-3i$ and its conjugate is:

    (A)   -5

    (B)   13

    (C)   14

    (D) $-5-6i$

    (E) $13-6i$

3. The expression $(2-5i)^2$ simplifies to:

    (A)   -21

    (B)   29

    (C) $-1-20i$

    (D) $-21-20i$

    (E) $-21-10i$

4. The expression $\dfrac{3}{3+2i}$ simplifies to:

    (A) $\dfrac{9-2i}{13}$

    (B) $\dfrac{9-6i}{13}$

    (C) $\dfrac{9-2i}{6}$

    (D) $\dfrac{3-2i}{2}$

    (E) $\dfrac{9-6i}{5}$

5. The expression $\dfrac{8-3i}{2-5i}$ simplifies to:

    (A) $\dfrac{31}{-21}$

    (B) $\dfrac{31+34i}{-21}$

    (C) $\dfrac{1-34i}{29}$

    (D) $\dfrac{1+34i}{29}$

    (E) $\dfrac{31+34i}{29}$

6. The value of i^8 is:

    (A) -1

    (B)   1

    (C) $-i$

    (D) i

7. The value of i^{886} is:

    (A) -1

    (B)   1

    (C) $-i$

    (D) i

8. The value of i^{21} is:

    (A) -1

    (B)   1

    (C) $-i$

    (D) i

9. The value of i^{355} is:

    (A) -1

    (B)   1

    (C) $-i$

    (D) i

Solutions to problems on complex numbers (unit 7.7)

1. (A)

$$(2-3i)+(2+3i)=2+2-3i+3i=4$$

2. (B)

$$(2-3i)(2+3i)=4+6i-6i-9i^2$$
$$=4-9(-1)=13$$

3. (D)

$$(2-5i)^2 = (2-5i)(2-5i)$$
$$= 4-10i-10i+25i^2 = -21-20i$$

4. (B)

$$\frac{3}{3+2i}=\left(\frac{3}{3+2i}\right)\left(\frac{3-2i}{3-2i}\right)=\frac{9-6i}{9-4i^2}=\frac{9-6i}{13}$$

5. (E)

$$\frac{8-3i}{2-5i}=\left(\frac{8-3i}{2-5i}\right)\left(\frac{2+5i}{2+5i}\right)=\frac{16-6i+40i-15i^2}{4-25i^2}$$
$$=\frac{31+34i}{29}$$

6. (B)

$$i^8=\left(i^2\right)^4=(-1)^4=1$$

7. (A)

$$i^{886}=\left(i^2\right)^{443}=(-1)^{443}=-1$$

8. (D)

$$i^{21}=i^{20}\cdot i=\left(i^2\right)^{10}i=(-1)^{10}i=i$$

9. (C)

$$i^{355}=i^{354}\cdot i=\left(i^2\right)^{177}i=-i$$

Unit 7.8 More geometry

The ACT® and SAT® subject tests tack on some geometry that is over and above what is needed for the SAT®.

Triangles

There are just a few more items on triangles. Consider the general triangle shown below.

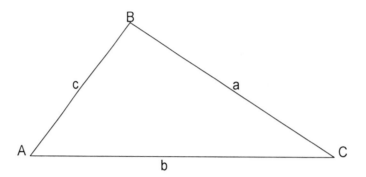

Sometimes you are expected to know something about an angle, based on the sides. In fact

$$\text{If } a^2 + b^2 = c^2 \text{ then } \angle C = 90°$$

$$\text{If } c^2 > a^2 + b^2 \text{ then } \angle C > 90° \text{ (obtuse)}$$

$$\text{If } c^2 < a^2 + b^2 \text{ then } \angle C < 90° \text{ (acute)}$$

If you think about this it makes sense. Start with a right triangle with angle C=90 degrees. As the hypotenuse (leg c) gets longer, angle C becomes larger (obtuse). As the hypotenuse (leg c) gets shorter, angle C becomes smaller (acute).

********************* ACT® and SAT® level 2 subject test only*******************************

NOTE: If you are taking the ACT® or SAT® level 2 subject test, you are expected to know the alternative formulas for the area of a triangle. Of course you already know that the area is $\frac{1}{2}bh$. But sometimes the base or height are not known. If you have a SAS situation (two sides and included angle are known), then these equivalent formulas can be applied:

$$A = \frac{1}{2}ab\sin C$$

$$A = \frac{1}{2}ac\sin B$$

$$A = \frac{1}{2}bc\sin A$$

Related to the formula for a triangle is the formula for the area of a parallelogram. The usual formula is base times height. But if you do not have the height, you can use

$$Area = ab\sin\theta$$

where a and b are any two adjacent sides of the parallelogram and θ is the angle between them (the SAS situation).

*********************** **End ACT ® and SAT® level 2 subject test material** *******************

You are also expected to know the formula for the area of an equilateral triangle with sides equal to s:

$$A = \frac{s^2\sqrt{3}}{4}$$

The proof is instructive because it involves a situation that is extremely common on tests. Also, if you have not memorized the formula above you may need to derive it. Suppose you are given that a triangle is equilateral and you know that its base is s but you do not have its height.

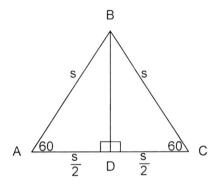

The height can be found by drawing the altitude to the base. Now we have two congruent right triangles by HL (the hypotenuses $\overline{AB}$ and $\overline{BC}$ are congruent because $\triangle ABC$ is equilateral, and the leg $\overline{BD}$ is congruent to itself). Therefore $\overline{AD}$ and $\overline{DC}$ are congruent. Each of the right triangles is 30-60-90, so the height must be $\frac{s\sqrt{3}}{2}$ because it is opposite the 60 degree angle. So the area of the equilateral triangle is

$$A = \frac{1}{2}bh = \left(\frac{1}{2}\right)s\left(\frac{s\sqrt{3}}{2}\right) = \frac{s^2\sqrt{3}}{4}$$

Trapezoid

You are expected to know that the area of a trapezoid is $\frac{1}{2}h(b_1 + b_2)$. The proof is instructive because it uses a common technique to analyze trapezoids.

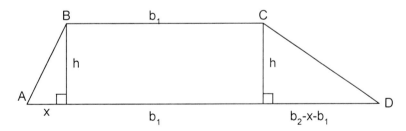

Use altitudes to divide the trapezoid into two right triangles and a rectangle. Compute the area of each figure and add them up to find the area of the trapezoid.

$$A = \frac{1}{2}(xh) + b_1h + \frac{1}{2}(b_2 - x - b_1)h = b_1h + \frac{1}{2}b_2h - \frac{1}{2}b_1h = \frac{1}{2}h(b_1 + b_2)$$

Circles

The most important additional information to be remembered about circles concerns the tangent line to a circle. The radius is perpendicular to the tangent line at the point of tangency. This gives rise to two congruent right triangles, when tangents are drawn from the same external point B as shown below ($\Delta OAB \cong \Delta OCB$).

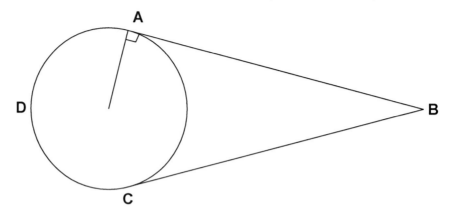

In this very common problem setup, note that

$$\overline{AB} \cong \overline{BC}$$

$$\angle B = \frac{1}{2}(arcADC - arcAC)$$

$$\angle B + arcAC = 180^0$$

You must also know that the equation of a circle with center (h, k) and radius r is:

$$\left(x-h\right)^2 + \left(y-k\right)^2 = r^2$$

This is an extension of the more common form of the circle whose center is at the origin:

$$x^2 + y^2 = r^2$$

Solids

The cylinder is a popular shape. Think of a soup can (or beer can if you prefer), with two circular ends. The volume of the cylinder is

$$V = \pi r^2 h$$

In general the volume of any prism is βh, where β is the area of the base and h is its height.

If there are other volume problems on the test, formulas will typically be supplied.

The surface area of a cylinder consists of the areas of the two circular ends plus the area of the side. To get the side, imagine cutting the side vertically and then flatten it out. This would yield a rectangular shape, with the width of h and the length equaling the circumference of the circle:

$$SA = 2\pi rh + 2\pi r^2$$

******************************* SAT® level 2 subject test only *********************************

Ellipse

The general equation of an ellipse with center (h, k) is

$$\frac{(x-h)^2}{a^2}+\frac{(y-k)^2}{b^2}=1$$

The ellipse will be horizontal (its major axis parallel to the x-axis) when $a^2 > b^2$. Otherwise the ellipse will be vertical. If you get stuck on these, it helps to find the intercepts. Substitute zero for x and then solve for y to find the y-intercepts. Substitute zero for y and solve for x to find the x-intercepts. Once the intercepts are known, it is easy to sketch the ellipse.

Hyperbola

The general equation of a hyperbola centered around the point (h, k) is

$$\frac{(x-h)^2}{a^2}-\frac{(y-k)^2}{b^2}=1$$

Notice that the equation for the ellipse is almost the same as the equation for the hyperbola, except with the ellipse the signs of the coefficients are both positive, whereas for the hyperbola the signs are positive and negative.

When the sign of the x-coefficient is positive (as shown above), the hyperbola is horizontal (it opens left and right). Find the x-intercepts by substituting zero for y and solving for x.

When the sign of the x-coefficient is negative (as shown below), the parabola is vertical (it opens up and down). Find the y-intercepts by substituting zero for x and solving for y.

$$\frac{(y-k)^2}{a^2}-\frac{(x-h)^2}{b^2}=1$$

************************* End SAT® level 2 subject test material *****************************

Problems on more geometry (unit 7.8)

1. The sides of three different triangles are given below. The triangle whose largest angle is acute is:

    I.   10, 20, 29

    II. 10, 15, 17

    III. 10, 24, 26

- (A)   Triangle I only
- (B)   Triangle II only
- (C)   Triangle III only
- (D)   Triangles I and II only
- (E)   Triangles I and III only

2. If a triangle has sides of 2 units, 3 units and 5 units, and its smallest angle is 25 degrees, what is its area?

- (A)   -1.99 square units
- (B)   -0.99 square units
- (C)   3.17 square units
- (D)   6.34 square units
- (E)   6.80 square units

3. The area of an equilateral triangle whose sides are of length 5 units is:

- (A) $1.25\sqrt{3}$ square units
- (B) $2.5\sqrt{3}$ square units
- (C) $5\sqrt{3}$ square units
- (D) $6.25\sqrt{3}$ square units
- (E) $12.5\sqrt{3}$ square units

4. A trapezoid has legs of 5 units each, and bases of 10 units and 16 units. Its area is:

- (A)   39 square units
- (B)   52 square units
- (C)   65 square units
- (D)   104 square units
- (E)   Cannot be determined

5. A tangent line is drawn from a point that is 10 units away from the center of a circle with a radius of 5 units. The length of the tangent line is:

- (A) $5\sqrt{3}$ units
- (B) $2\sqrt{5}$ units
- (C) $3\sqrt{5}$ units
- (D) $4\sqrt{5}$ units
- (E) $5\sqrt{5}$ units

6. Two lines tangent to a circle meet at an external point, forming a 30 degree angle between them. The lines form a minor arc on the circle. The measure of the minor arc is:

- (A)   30degrees
- (B)   60degrees
- (C)   90degrees
- (D)   150degrees
- (E)   330degrees

7. If a circle is tangent to the x-axis at x=5 and is tangent to the y-axis at y=-5, what is the equation of the circle?

- (A) $x^2 + y^2 = 25$
- (B) $(x+5)^2 + (y+5)^2 = 25$
- (C) $(x+5)^2 + (y-5)^2 = 25$
- (D) $(x-5)^2 + (y+5)^2 = 25$
- (E) $(x-5)^2 + (y-5)^2 = 25$

8. A cylinder has a volume of 100π cubed units and a radius of 5 units. What is its height?

- (A)   2 units
- (B)   4 units
- (C)   8 units
- (D)   10 units
- (E)   20 units

9. <u>SAT® level 2 only.</u> An ellipse with the equation $4x^2 - 16x + 9y^2 = 20$ has its center at:

- (A)   (0, 0)
- (B)   (-2, 0)
- (C)   (0, -2)
- (D)   (2, 0)
- (E)   (0, 2)

Solutions to problems on more geometry (unit 7.8)

1. (B)

In Triangle I, the largest obtuse is obtuse because
$29^2 > 10^2 + 20^2$.

In Triangle II, the largest angle is acute because
$17^2 < 10^2 + 15^2$.

In Triangle III, the largest angle is a right angle
because $10^2 + 24^2 = 26^2$

2. (C)

The smallest angle is opposite the smallest side. That
means that the 25 degree angle must be opposite the
side of length 2 and it must be enclosed by the sides
of lengths 3 and 5. The area of the triangle is:

$$A = \frac{1}{2}ab\sin\theta = \frac{1}{2}(3)(5)\sin 25 = 3.17$$

3. (D)

$$A = \frac{s^2\sqrt{3}}{4} = \frac{25\sqrt{3}}{4} = 6.25\sqrt{3}$$

4. (B)

If you sketch the trapezoid, you will find that its
altitudes form two 3-4-5 right triangles. Therefore
the height is 4 units.

$$A = \frac{1}{2}h(b_1 + b_2) = \frac{1}{2}4(10+16) = 52$$

5. (A)

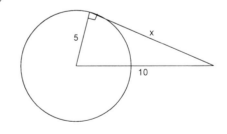

$$5^2 + x^2 = 10^2$$
$$x = \sqrt{75} = 5\sqrt{3}$$

6. (D)

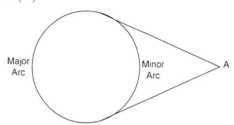

Angle A + minor arc=180

minor arc=180-30=150

7. (D)

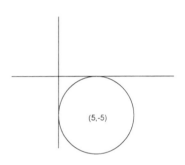

center=$(5,-5)$, radius=5

$$(x\text{-}5)^2 + (y+5)^2 = 5^2$$

8. (B)

$$100\pi = \pi 5^2 h, \ h = 4$$

9. (D)

Complete the square:

$$4x^2 - 16x + 9y^2 = 20$$
$$4(x^2 - 4x) + 9y^2 = 20$$
$$4(x^2 - 4x + 4) + 9y^2 = 20 + 16$$
$$4(x-2)^2 + 9y^2 = 36$$
$$\frac{(x-2)^2}{9} + \frac{y^2}{4} = 1$$

Unit 7.9 Trigonometry for Right Triangles

The ACT® and SAT® subject tests have a few trigonometry problems. Many of them are easy if you know the basic information that is reviewed here. It is not necessary to be a trigonometry wizard to do well on the test.

This unit contains the trigonometry usually taught in Geometry class. It is derived from the properties of right triangles. The main idea is captured in the infamous mnemonic, SOHCAHTOA. Take a look at this diagram:

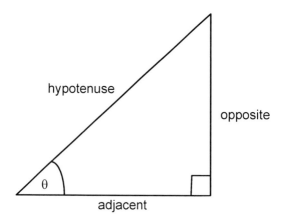

Notice that we have a right triangle, and we want to find the trigonometric ratios for one of its angles, labeled θ in the diagram. The ratios are:

S = $\sin \theta$ is equal to

O =   opposite over

H =   hypotenuse.

C = $\cos \theta$ is equal to

A =   adjacent over

H =   hypotenuse.

T = $\tan \theta$ is equal to

O =   opposite over

A =   adjacent.

From the SOHCAHTOA definitions, we can see that

$$\frac{\sin \theta}{\cos \theta} = \frac{\dfrac{opp}{hyp}}{\dfrac{adj}{hyp}} = \frac{opp}{adj} = \tan \theta$$

This is an important identity. The other really important identity is

$$\sin^2 \theta + \cos^2 \theta = 1$$

You are also expected to know the three reciprocal functions shown below.

$$\frac{1}{\sin\theta} = \csc\theta$$

$$\frac{1}{\cos\theta} = \sec\theta$$

$$\frac{1}{\tan\theta} = \cot\theta$$

<u>Calculator Tip</u>: Although the sine, cosine and tangent functions are on the graphing calculator, the reciprocal functions are not. They may be found getting a trigonometric value and then taking its inverse. For example, the $\csc 25°$ would be found by the keystrokes *SIN-25-)-enter-* x^{-1} *-enter.*

There are special angles, called reference angles, and you are expected to know the exact values of their trigonometric functions (the approximate values can be found on your graphing calculator, using the SIN, COS and TAN keys). Fortunately these can be derived easily from the special right triangles that were memorized earlier for the SAT®.

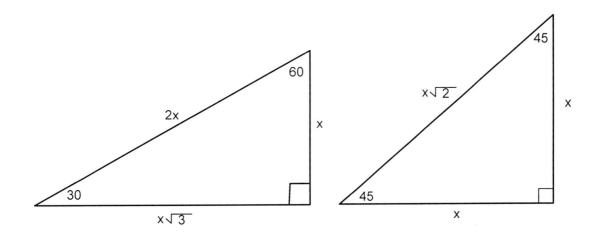

	$\sin\theta$	$\cos\theta$	$\tan\theta$
0	0	1	0
30	$\dfrac{x}{2x} = \dfrac{1}{2}$	$\dfrac{x\sqrt{3}}{2x} = \dfrac{\sqrt{3}}{2}$	$\dfrac{x}{x\sqrt{3}} = \dfrac{1}{\sqrt{3}} = \dfrac{\sqrt{3}}{3}$
45	$\dfrac{x}{x\sqrt{2}} = \dfrac{1}{\sqrt{2}} = \dfrac{\sqrt{2}}{2}$	$\dfrac{x}{x\sqrt{2}} = \dfrac{1}{\sqrt{2}} = \dfrac{\sqrt{2}}{2}$	$\dfrac{x}{x} = 1$
60	$\dfrac{x\sqrt{3}}{2x} = \dfrac{\sqrt{3}}{2}$	$\dfrac{x}{2x} = \dfrac{1}{2}$	$\dfrac{x\sqrt{3}}{x} = \sqrt{3}$
90	1	0	Undefined

Problems on trigonometry for right triangles (unit 7.9)

Use the figure below to solve problems 1-3.

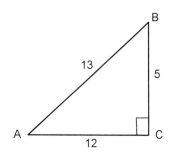

1. For the right triangle above, the $\sec A$ is:

   (A) $12/5$

   (B) $13/5$

   (C) $12/13$

   (D) $13/12$

   (E) $5/13$

2. For the triangle above, $\sin A + \cos B$ equals

   (A) $10/13$

   (B) 1

   (C) $17/13$

   (D) $24/13$

   (E) $30/13$

3. The following are equal to $5/12$ EXCEPT:

   (A) $\tan A$

   (B) $\cos B$

   (C) $\cot(90 - A)$

   (D) $1/\cot A$

   (E) $\cot B$

4. If the $\sin \theta = 5/7$ then the $\cot \theta$ equals:

   (A) $\dfrac{5}{12}$

   (B) $\dfrac{7}{12}$

   (C) $\dfrac{13}{12}$

   (D) $\dfrac{7}{5}$

   (E) $\dfrac{2\sqrt{6}}{5}$

5. The expression $\dfrac{\sin^2 \theta - \sin^4 \theta}{\cos^4 \theta}$ simplifies to:

   (A) $\sin^2 \theta$

   (B) $\cos^2 \theta$

   (C) $\tan^2 \theta$

   (D) $\tan^4 \theta$

   (E) $2 \sin \theta \cos \theta$

6. The $(\cos 30°)(\sin 45°)$ is equal to:

   (A) $\dfrac{\sqrt{5}}{2}$

   (B) $\dfrac{\sqrt{5}}{4}$

   (C) $\dfrac{\sqrt{6}}{2}$

   (D) $\dfrac{\sqrt{6}}{4}$

   (E) $\dfrac{\sqrt{2}}{4}$

7. If the ratio of the angles in a triangle is 3:2:1 then the sine of the smallest angle is:

   (A) $\dfrac{1}{2}$

   (B) $\dfrac{\sqrt{2}}{4}$

   (C) $\dfrac{\sqrt{3}}{2}$

   (D) 1

   (E) $\dfrac{\sqrt{3}}{3}$

Solutions to problems on trigonometry for right triangles (unit 7.9)

1 (D)

$$\cos A = \frac{12}{13}$$

$$\sec A = \frac{1}{\cos A} = \frac{13}{12}$$

2. (A)

$$\sin A + \cos B = \frac{5}{13} + \frac{5}{13} = \frac{10}{13}$$

3. (B)

$$\cos B = \frac{5}{13}$$

4. (E)

If a right triangle has a hypotenuse of 7 and one of its legs is 5, then the length of the other leg is:

$$\sqrt{7^2 - 5^2} = \sqrt{49 - 25} = \sqrt{24} = 2\sqrt{6}$$

$$\cot \theta = \frac{2\sqrt{6}}{5}$$

5. (C)

$$\frac{\sin^2 \theta - \sin^4 \theta}{\cos^4 \theta} = \left(\frac{\sin^2 \theta}{\cos^2 \theta} \right) \frac{(1 - \sin^2 \theta)}{\cos^2 \theta}$$

$$= \tan^2 \theta \left(\frac{\cos^2 \theta}{\cos^2 \theta} \right) = \tan^2 \theta$$

6. (D)

$$(\cos 30°)(\sin 45°) = \frac{\sqrt{3}}{2} \cdot \frac{\sqrt{2}}{2} = \frac{\sqrt{6}}{4}$$

7. (A)

This is a 30-60-90 triangle. It can be found using the approach to mixture problems from unit 3.6:

$$3x + 2x + x = 180, \ 6x = 180, \ x = 30$$

$$\sin 30 = \frac{1}{2}$$

Unit 7.10 Trigonometry for all triangles

Skip this unit if your only requirement is to prepare for the SAT® Level 1 subject test. Otherwise, if you are preparing for the ACT® or SAT® Level 2 subject test, you should cover this unit.

The trigonometry in the previous unit is usually taught in Geometry class and is usually applied in the context of right triangles. Later, in pre-calculus or analysis, you learn that the trigonometric functions apply to angles universally. The $\sin 30°$ is $1/2$ whether or not that thirty degree angle lies in a right triangle or some other kind of triangle. Don't worry if you have not taken pre-calculus or analysis – what you need to know is covered here.

Radian Measure

In addition to measuring angles in degrees, it is necessary to know how to measure angles in radians. To convert between radians and degrees by hand, just remember that π radians is equal to 180 degrees. With this knowledge, you can set up a ratio to do any conversion

$$\frac{\pi}{180} = \frac{\text{radians}}{\text{degrees}}.$$

Suppose we want to convert 25 degrees to radians. We plug-in, cross-multiply and solve for radians:

$$\frac{\pi}{180} = \frac{r}{25} \; , \; 180r = 25\pi \; , \; r = \frac{25\pi}{180} = 0.436$$

If we wanted to convert 2 radians to degrees:

$$\frac{\pi}{180} = \frac{2}{d} \; , \; \pi d = 360 \; , \; d = \frac{360}{\pi} = 114.6$$

<u>Calculator Tip</u>: Conversions may also be done on the graphing calculator. To convert 25 degrees to radians, set the mode of the calculator to radians. Next enter 25 followed by 2nd-ANGLE-1. This gives 25 degrees. Hit the enter key. This gives you the answer in radians. To convert 2 radians to degrees, set the mode of the calculator to degrees. Next enter 2 followed by 2nd-ANGLE-2. This gives radians. Hit the enter key. This gives you the answer in degrees. When converting angles on the calculator, make sure the mode is set to the target (degrees or radians).

Any triangle

Previously, you learned the definition of trigonometry functions in the context of the right triangle. Also you know that $a^2 + b^2 = c^2$, but only when a and b are legs of a right triangle and when c is the hypotenuse of a right triangle. This section extends those ideas to any triangle, so this material is much more useful (in the real world, most triangles are probably not right triangles).

Consider the general triangle shown below. Notice that it is not a right triangle, or any special triangle.

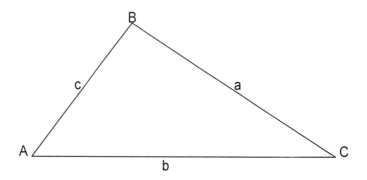

The vertices (angles) of the triangle are denoted by capital letters and the legs of the triangle are denoted by lower case letters.

There are two "laws" that are needed about triangles in general, and also may be applied to right triangles. The first is the law of sines,

$$\frac{\sin A}{a} = \frac{\sin B}{b} = \frac{\sin C}{c}$$

In words, the law of sines says that the ratio of the sine of an angle to the length of its opposite side is the same for all of the angles in any triangle.

The other is the law of cosines:

$$a^2 = b^2 + c^2 - 2bc \cos A$$

$$b^2 = a^2 + c^2 - 2ac \cos B$$

$$c^2 = a^2 + b^2 - 2ab \cos C$$

Each of the equations above is equivalent. The last equation should remind you of the Pythagorean Theorem. If angle C is a right angle, cosC is zero, and the law of cosines becomes the same as the Pythagorean Theorem.

The law of sines is most often applied in AAS, ASA and SSA situations. You are given some information and expected to derive the rest using the law of sines. For example, with AAS suppose you are given that A=25, B=40 and a=10. To find b

$$\frac{\sin 25}{10} = \frac{\sin 40}{b} \, , \quad b = \frac{10 \sin 40}{\sin 25} = \frac{10(0.6428)}{0.4226} = 15.21$$

The law of cosines is most often applied in SAS and SSS situations. For example, with SAS suppose you are given that b=10, c=15 and A=30. To find a:

$$a^2 = 10^2 + 15^2 - 2(10)(15)\cos 30 = 100 + 225 - 300\left(\frac{\sqrt{3}}{2}\right) = 65.2$$

Though they are not absolutely essential, there are some other formulas from trigonometry that can be useful on the the SAT® Level 2 subject test. Two other identities are:

$$\sec^2 \theta - \tan^2 \theta = 1$$

$$\csc^2 \theta - \cot^2 \theta = 1$$

The two most common double angle formulas are:

$$\sin 2x = 2 \sin x \cos x$$

$$\cos 2x = 1 - 2 \sin^2 x = 2 \cos^2 x - 1$$

Changes to amplitude and period

	Amplitude	Period	Phase Shift	Vertical Shift
$y = a\sin b(x-h)+k$	$\lvert a \rvert$	$\dfrac{2\pi}{\lvert b \rvert}$	h	k
$y = a\cos b(x-h)+k$	$\lvert a \rvert$	$\dfrac{2\pi}{\lvert b \rvert}$	h	k
$y = a\tan b(x-h)+k$	undefined	$\dfrac{\pi}{\lvert b \rvert}$	h	k

The table above summarizes the effect of translations (sliding a graph horizontally or vertically) and scale changes (multiplying either variable by a constant) on the graphs of the trigonometric functions.

Sliding the graph horizontally by h units is called a phase shift. Sliding the graph vertically by k units is called a vertical shift.

Multiplying the y-variable by a constant causes the amplitude of the graph to change. Recall that the amplitude is ½ of the difference between the maximum and minimum values of the function. Multiplying the x-variable by a constant causes the period to change. The period is the length of a full repeat of the function.

The functions $\sin x$ and $\cos x$ have periods of 2π and amplitudes of 1. The function $\tan x$ has a period of π and an undefined amplitude (the tangent ranges from $-\infty$ to $+\infty$).

Problems on trigonometry for all triangles (unit 7.10)

1. What is the radian measure of an angle that is 150 degrees?

   (A) $-\pi/6$

   (B) $\pi/6$

   (C) $\pi/5$

   (D) $5\pi/6$

   (E) $6\pi/5$

2. What is the degree measure of an angle that is $\dfrac{7\pi}{6}$ radians?

   (A)   4

   (B)   154

   (C)   210

   (D)   240

   (E)   1260

3. In the triangle below, find the value of x:

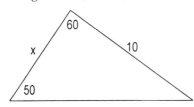

   (A)   8.2 units

   (B)   8.8 units

   (C)   10.8 units

   (D)   11.3 units

   (E)   12.3 units

4. In the triangle below, find the value of x:

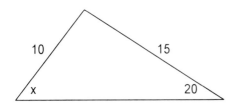

(A)   0.2 degrees

(B)   0.5 degrees

(C)   22.7 degrees

(D)   30.9 degrees

(E)   129.1 degrees

5. In the triangle below, find the value of x:

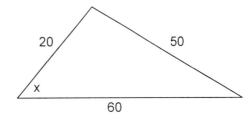

(A)   0.6 degrees

(B)   18.2 degrees

(C)   51.3 degrees

(D)   88.6 degrees

(E)   110.5 degrees

6. Compared to the original equation, $y = \cos x$, the amplitude and period of the new equation, $y = \cos(x-2)+2$, are:

   (A)   both doubled

   (B)   both halved

   (C)   amplitude doubled, period halved

   (D)   amplitude halved, period doubled

   (E)   both unchanged

7. Find the amplitude and period of the following equation: $y = 2\sin(3x-9)+7$.

   (A)   amplitude=3, period=4π

   (B)   amplitude=2, period=$2\pi/3$

   (C)   amplitude=1/2, period=$2\pi/3$

   (D)   amplitude=1/2, period=π

   (E)   amplitude=2, period=6π

Solutions to problems on trigonometry for all triangles (unit 7.10)

1. (D)

$$\frac{\pi}{180} = \frac{r}{150} \ , \ r = \frac{150\pi}{180} = \frac{5\pi}{6}$$

2. (C)

$$\frac{\pi}{180} = \frac{\frac{7\pi}{6}}{d} \ , \ \pi d = \frac{1260\pi}{6} \ , \ d = 210$$

3. (E)

The third angle must be 70 degrees:

$$180 - (50 + 60) = 70$$

$$\frac{\sin 50}{10} = \frac{\sin 70}{x} \ , \ x = \frac{10 \sin 70}{\sin 50} = 12.267$$

4. (D)

$$\frac{\sin 20}{10} = \frac{\sin x}{15} \ , \ \sin x = \frac{15 \sin 20}{10} = 0.513, x = 30.9$$

5. (C)

$$50^2 = 20^2 + 60^2 - 2(20)(60)\cos X$$

$$\cos X = 0.625, X = 51.3$$

6. (E)

This equation has been translated, and therefore the amplitude and period are unchanged. The curve has been shifted right by 2 units and up by 2 units.

7. (B)

$$y = 2\sin(3x - 9) + 7 = 2\sin 3(x - 3) + 7$$

$$a = 2, \ b = 3$$

amplitude = 2, period = $2\pi / 3$

7.11 Miscellaneous topics for the SAT® Level 2 subject test

Below are some notes on miscellaneous topics for the SAT® level 2 subject test. These topics do not occur as frequently as the other previous topics, but they do occur. If you are aiming for a high score (700+), you should be familiar with these topics.

Vectors

Let $\mathbf{u} = (u_1, u_2)$ and $\mathbf{v} = (v_1, v_2)$. Then

$$k\mathbf{u} = (ku_1, ku_2)$$ scalar multiplication

$$\|\mathbf{u}\| = \sqrt{u_1^2 + u_2^2}$$ magnitude (vector length)

$$\mathbf{u} \pm \mathbf{v} = (u_1 \pm v_1, u_2 \pm v_2)$$ addition/subtraction

$$\mathbf{u} \cdot \mathbf{v} = u_1 v_1 + u_2 v_2$$ dot product

$$\cos\theta = \frac{\mathbf{u} \cdot \mathbf{v}}{\|\mathbf{u}\|\|\mathbf{v}\|}$$ θ is the angle between $\mathbf{u}$ and $\mathbf{v}$.

Polar Coordinates

The location of a point in a plane is usually given using the Cartesian coordinate system (x, y). But the location of a point may also be determined using polar coordinates, (r, θ), where r is the length of the line segment connecting the point with the origin, and θ is the angle formed between the line segment and the horizontal axis.

The relationships between these two coordinate systems are illustrated below. Note that $x = r\cos\theta$, $y = r\sin\theta$, and $r^2 = x^2 + y^2$.

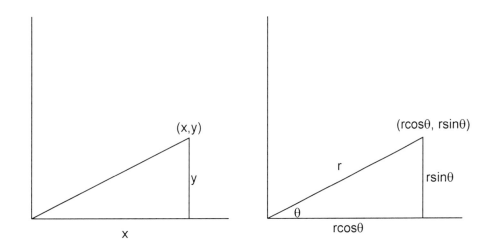

Problems on miscellaneous topics for the SAT® level 2 subject test (unit 7.11)

1. Find the magnitude of the vector that begins at the origin and terminates at the point (-3, 5).

(A)   2

(B)   3

(C)   4

(D)   6

(E) $\sqrt{34}$

2. Find the polar coordinates of the point with the Cartesian coordinates of $(4\sqrt{3}, -4)$.

(A) $(4, 30°)$

(B) $(6, 30°)$

(C) $(8, 30°)$

(D) $(4, 330°)$

(E) $(8, 330°)$

3. Find the Cartesian coordinates of the point with the polar coordinates of $(\sqrt{8}, 45°)$.

(A) $(2, 2)$

(B) $(2, \sqrt{8})$

(C) $(\sqrt{8}, 2)$

(D) $(4, 4)$

(E) $(2\sqrt{6}, 2\sqrt{6})$

4. Two vectors begin at the origin and end at the coordinates of (3,-1) and (2, 10). What are the coordinates of the sum of these two vectors?

(A)   (1, 9)

(B)   (1, -11)

(C)   (5, 9)

(D)   (5, 11)

(E)   (-1, 11)

5. Two vectors begin at the origin and end at the coordinates of (3,-1) and (2, 10). What is the dot product of these two vectors?

(A)   -16

(B)   -6

(C)   -4

(D)   4

(E)   16

6. Two vectors begin at the origin and end at the coordinates of (3,-1) and (2, 10). How large is the angle between these two vectors?

(A)   7.1 degrees

(B)   82.9 degrees

(C)   86.2 degrees

(D)   97.1 degrees

(E)   262.9 degrees

7. If two vectors are orthogonal (at right angles with each other), their dot product must be:

(A)   -1

(B)   0

(C)   1

(D)   infinite

(E)   cannot be determined

Solutions to problems for the SAT® level 2 subject test (unit 7.11)

1. (E)

$$\sqrt{(-3)^2 + 5^2} = \sqrt{9 + 25} = \sqrt{34}$$

2. (E)

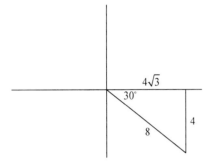

The coordinates $(4\sqrt{3}, -4)$ form a 30-60-90 triangle with a hypotenuse of 8. The reference angle is 30 degrees. So the polar coordinates are (8, 330).

3. (A)

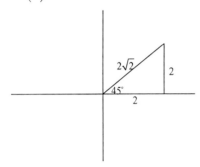

This forms a 45-45-90 triangle, with a hypotenuse of $\sqrt{8} = 2\sqrt{2}$ and legs of length 2.

4. (C)

$$(3, -1) + (2, 10) = (3 + 2, 10 - 1) = (5, 9)$$

5. (C)

$$(3, -1) \cdot (2, 10) = 3(2) - 1(10) = 6 - 10 = -4$$

6. (D)

$$\|\mathbf{u}\| = \sqrt{3^2 + (-1)^2} = \sqrt{10}$$
$$\|\mathbf{v}\| = \sqrt{2^2 + 10^2} = \sqrt{104} = 2\sqrt{26}$$

In problem 5 we found that the dot product is -4. Next find:

$$\cos\theta = \frac{\mathbf{u} \cdot \mathbf{v}}{\|\mathbf{u}\|\|\mathbf{v}\|} = \frac{-4}{\sqrt{10} \cdot 2\sqrt{26}}$$

$$= \frac{-2}{\sqrt{260}} = \frac{-2}{2\sqrt{65}} = \frac{-\sqrt{65}}{65}$$

$$\theta = 97.1°$$

7. (B)

If two vectors are orthogonal then the angle between them is 90 degrees. Substituting, we get

$$\cos 90 = 0 = \frac{\mathbf{u} \cdot \mathbf{v}}{\|\mathbf{u}\|\|\mathbf{v}\|}$$

Therefore the dot product is zero.

Breinigsville, PA USA
21 April 2010
236627BV00001B/6/P